"*Inviting Disaster* is a sobering and instructive book. It is full of gripping stories of what can go wrong when modern technology is not treated with proper attention and respect."
> —Henry Petroski, author of *To Engineer Is Human*

"This book belongs on the shelf of anyone who has felt the call to test man's delicate relationship with his machines. James Chiles has done a great service by chronicling the lowest points in our brief history with our most useful—and most dangerous—inventions. He reminds us to remain vigilant in our zeal, train for the worst, heed the warning signs of impending doom, and trust our intuition to get out of bad situations. Engineers and executives, system designers and supervisors would do well to incorporate the lessons from the many cases offered here."
> —Dr. Gary Klein, author of *Sources of Power: How People Make Decisions*

"A cornucopia of disasters, sharply honed, terrifying, yet gracefully and engagingly presented. Even the familiar ones are freshly revealed."
> —Charles Perrow, author of *Normal Accidents*

"This book offers a thorough examination of industrial failure and its causes. A look into some of technology's most miserable moments. Instructive stuff."
> —Gerald C. Meyers, crisis management consultant, former chairman and CEO of American Motors Corporation

"*Inviting Disaster* is a rare book: a thoughtful, analytical page-turner. From the *Challenger* to Three Mile Island, from dam failures to building collapses, James Chiles brings alive the technological choices that undergird modern society. He shows how easily they can go disastrously wrong—and how we might fix the problem."
> —Steven Lubar, chair, Division of the History of Technology, Smithsonian National Museum of American History

"The lessons are clear and disconcerting: a sequence of minor human errors when combined with elementary technical problems can lead to unimaginable catastrophe."
> —Kenneth L. Carper, professor, Washington State University, and editor, ASCE *Journal of Performance of Constructed Facilities*

"Surrounded by a growing sense of loss of control over our creations, we desperately need to discover and learn from those people who have found successful strategies for controlling our hazardous tools. Technology writer James Chiles has found these people . . . and he is the human bridge between us and those who know what we need to know to operate safely in the future."

—James Oberg, veteran space engineer, author of *Uncovering Soviet Disasters* and *Star-Crossed Orbits: Inside the U.S./Russian Space Alliance*

"In his interesting, well-written, and readable book James Chiles analyzes a number of disasters and near-misses and shows how they could have been avoided—and how we can avoid similar events in the future."

—Trevor Kletz, process safety consultant, author of *By Accident: A Life Preventing Them in Industry*

"Curiosity is the spur that urges us to wander in new realms; writers like Chiles make it rewarding to yield to that innate wanderlust."

—George C. Larson, editor, *Air & Space Magazine*

INVITING DISASTER

INVITING DISASTER

LESSONS FROM THE EDGE OF TECHNOLOGY

AN INSIDE LOOK AT CATASTROPHES ——————

—————————————— AND WHY THEY HAPPEN

JAMES R. CHILES

HarperBusiness

An Imprint of HarperCollins*Publishers*

Portions of chapter 9 appeared in "High Tension," *Air & Space/Smithsonian*
(March 2001), by the author.

HarperCollins books may be purchased for educational, business, or sales
promotional use. For information please write: Special Markets Department,
HarperCollins Publishers, Inc., 10 East 53rd Street, New York, NY 10022.

FIRST EDITION

Designed by William Ruoto

Library of Congress Cataloging-in-Publication Data

Chiles, James R.
Inviting disaster / James R. Chiles—1st ed.
p. cm.
ISBN 0-06-662081-3
1. Technology—Risk assessment. I. Title.
T174.5.C57 2001
363.1—dc21   2001024074

01 02 03 04 05   **QPF**   10 9 8 7 6 5 4 3 2 1

To Dick and Ellen,
children of the Great Depression:

I've never known people who worked so hard
and had so much fun.

*"The horror of that moment," the King went on,
"I shall never, never forget."*

*"You will, though," the Queen said, "if you
don't make a memorandum of it."*

—Lewis Carroll, *Through the Looking Glass*
(1871)

CONTENTS

ACKNOWLEDGMENTS xiii

INTRODUCTION 1
  ON THE MACHINE FRONTIER: NEW TECHNOLOGY AND
  OLD HABITS

1: **SHOCK WAVE** 17
  HIGH TECH ON THE HIGH SEAS

2: **BLIND SPOT** 37
  BAFFLED AND BEWILDERED INSIDE THE MASSIVE
  SYSTEM

3: **RUSH TO JUDGMENT** 65
  WHEN FLAGSHIP PROJECTS RUN OUT OF TIME

4: **DOUBTLESS** 95
  TESTING IS SUCH A BOTHER

5: **THE REALLY BAD DAY** 117
  PANIC AND TRIUMPH ON THE MACHINE FRONTIER

6: **TUNNEL VISION   139**
    GO AWAY, I'M BUSY

7: **RED LINE RUNNING   161**
    HUMANS HAVE A LIMIT, TOO

8: **A CRACK IN THE SYSTEM   181**
    FAILURE STARTS SLOW, BUT IT GROWS

9: **THE HEALTHY FEAR   205**
    ALIVE AND ALERT AT DANGER'S EDGE

10: **THAT HUMAN TOUCH   229**
     HOW LITTLE ERRORS MAKE BIG ACCIDENTS

11: **ROBBING THE PILLAR   259**
     SLACKING OFF WITH THE HIGH-POWER SYSTEM

12: **MACHINE MAN   275**
     SURVIVING AND THRIVING ON THE NEW FRONTIER

**DISASTERS, CALAMITIES, AND NEAR MISSES
CITED IN THE BOOK   295**

LIST OF KEY SOURCES   313

INDEX   331

ACKNOWLEDGMENTS

Family. My love to Chris for the indomitable way she held the family fort during the time it took me to bring back the story. My three boys and parents-in-law cheerfully bore all adversities, from grouchiness to papers stacked in all the wrong places. I can't begin to thank my brothers, M. C. and Dan, for the friendship and encouragement they offered at every turn, from our chemistry-set days to the present.

East Coast people. My appreciation to Jay Mandel of the William Morris Agency for expert help in tossing that long ball downfield. Equal thanks to Dave Conti and Amanda Maciel of HarperBusiness for catching it. Your patience and humor brought me through. In 1983 Don Moser and Jack Wiley of *Smithsonian* gave a new writer a chance to pursue a crazy idea about how great structures would age if abandoned to the eons. Thanks for letting my nose in the tent.

Bosses. My employers at the state of Minnesota have put up with some short-notice vacation slips over the last ten years when I had to chase a story. Thank you for the support. Writing is fun but I value that day job, too.

The real world. My hat is off to all employees who do the tough and risky work while the rest of us sleep, and to world-class engineers, like Les Robertson, who make sure our great structures will stand up for as long as we want to take care of them. My gratitude goes out to the companies and agencies that let me inside to be a fly on the wall. These include Haverfield and its high-voltage crew; Dyno Nobel; Blue Goose Drilling; the New York Water Tunnel No. 3 Project; Butch Strickland and his tall-tower boys; and the Los Angeles County Fire Department. You taught me that a little fear goes a good, long way.

INTRODUCTION

ON THE MACHINE FRONTIER: NEW TECHNOLOGY AND OLD HABITS

To see what kind of strange new world we are building for ourselves, consider what happened in January 1969 at the Hungarian Carbonic Acid Producing Company, at Répcelak, Hungary. The company was in the business of removing CO_2 from natural gas and selling it. The liquid was stored in small cylinders as well as in four big storage tanks, cooled by ammonia refrigeration. The gas arrived at the plant with traces of water in it that had to be removed. On occasion this stray water caused gauges, fittings, level indicators, and even safety valves to freeze shut. But the plant kept running.

On December 31, 1968, the plant shut down with the indicators showing at least twenty tons of liquid CO_2 in each tank. The plant opened again late on the night of January 1. Running short of cylinders to store the liquid CO_2, operators directed the flow into storage tank C, which was supposed to have plenty of capacity. About a half hour later tank C exploded, and its fragments blew apart tank D.

The twin explosions killed four people nearby and ripped tank A from its foundation bolts, tearing a hole about a foot across. In escaping furiously through the new opening, the pressurized, liquid CO_2 acted like a rocket propellant. Tank A took off under the thrust, crashing through a wall into the plant laboratory, dumping out tons of liquid CO_2 across the floor, and instantly freezing five people where they stood. The deluge left the room at a temperature of $-108°F$, starved of breathable air, and covered with a thick layer of dry ice.

We have been hard at work for more than two centuries now, building a world out of cold iron that is very far from our ancient instincts and traditions, and becoming more so. Machines going crazy are among the few things left on this civilized planet that can still inspire deep dread. I mean the kind of dread that railroad foreman James Roberts felt one wild night on December 28, 1879, when he ventured out onto the mile-long Tay Bridge, crossing a bay off eastern Scotland.

He was looking for a train that had rolled into the darkness to cross the bridge but had not reported in from the other side. With storm winds so high that he had to crawl a third of a mile along the bridge on his hands and knees, he stopped at a new chasm, opening onto the black waters eighty-eight feet below. A third of the bridge had collapsed into the Tay River estuary, taking the entire train and seventy-five passengers with it.

That bridge fell as a result of a combination of design errors and quality control problems, exposed by the high winds and the train's passage. Those kinds of problems continue, but the consequences are higher. Each year the margins of safety draw thinner, and the energies that we harness grow in power. The specs of our equipment may surprise you. Petrochemical plants have pressure vessels operating at twenty thousand pounds per square inch; modern coal-fired power plants have combustion chambers so big that an eight-story office building would fit easily inside the furnace of some of these monsters. Pulverized coal shoots into their combustion chambers, making a roiling, continuous fireball in the center.

In the cause of cost cutting, our machines keep getting bigger, putting more eggs in fewer baskets. The new Airbus A380 double-decker jetliner will start with 555 seats but has the capacity to eventually carry eight hundred people, putting potential death tolls into the passenger-ship category. And marine insurers are vexed about a proposed new generation of giant container-carrying ships. The biggest container carriers now fit only 3,500 full-size cargo boxes; the new ones should fit up to 10,000 of the forty-foot boxes. A single such ship if lost at sea with all cargo could sock underwriters with a loss of $2 billion or more.

The most awesome machines working today are not easily viewed because they are either kept in no-trespassing zones or used in remote locales. Recently television viewers were surprised by footage of the 505-foot, 8,300-ton destroyer USS *Cole* being carried piggyback on the heavy-lift vessel *Blue Marlin*. The *Blue Marlin*'s earlier work had been out of the media spotlight, hauling rigs and equipment for off-shore oil fields.

Our machines take us into risky locales, which might be outer space, up on a two-thousand-foot-tall tower, or on an artificial island, making our lives entirely dependent on their proper functioning. A mile-long complex of drilling platforms and petroleum processing plants called Ekofisk sits in the stormy North Sea, far from view of shore. Workers excavate salt mines far under great bodies of water; one of these mines drained a thousand-acre lake in Louisiana in 1980, after a drilling rig punched a hole in the mine's roof.

Some industries, such as nuclear power and chemical processing plants, have been operating more cautiously after infamous disasters in the 1980s, but others have taken their place in the headlines. The last two years have seen the twin failures of *Mars Climate Orbiter* and *Mars Polar Lander*; an unintentional nuclear reaction at the JCO Tokai Works Conversion Test Facility in Japan; and a rash of fires and explosions at fossil-fuel power plants nationwide in 1999. In June 1995 the *Royal Majesty* cruise ship grounded on the shoals near Nantucket Island because the cable to its Global Positioning System (GPS) antenna had come loose. Nobody on the ship's bridge noticed that the ship was miles off course. Normally the depth alarm would have gone off when less than ten feet of water remained under the hull, but somebody had set the alarm to stay quiet until zero feet of water remained. Such chains of error and mishap events occur throughout our modern world.

Though the size and power of our machines have risen enormously, it still doesn't take much to trigger a disaster. The catalyst for the crash of an Air France Concorde in July 2000 was a mere strip of titanium that had fallen onto the runway from the engine of a DC-10 minutes before. The strip was only eighteen inches long, but when the main landing gear for the supersonic transport hit it on takeoff, a tire blew out (see Figure 1). According to French investigators, the exploding tire launched a ten-pound slab of rubber, which slammed against the underside of a wing tank called "Fuel Collector Tank No. 1."

The impact caused shock waves throughout the tank. These focused to blow out a big section of the tank wall from the *inside*. As fuel gushed out of the bottom at the rate of twenty-six gallons a second, the kerosene went into the engine air intakes on the left side, which scooped up the fuel along with air the engines needed. The leaking fuel caught fire, and all this combustion in the wrong place caused the left engines to lose power. Pilots Christian Marty and Jean Marcot saw a fire alarm on the number 2 engine and shut it down. It was actually a false alert, because the engines' fuel systems were not on fire. Still, both left-hand engines looked in severe distress, trailing flames more than two hundred feet to the rear as the Concorde left Charles de Gaulle Airport. The pilots held the nose up steeply to try and keep the transport flying until it could make an emergency landing at Le Bourget. But after less than two minutes in the air the Concorde rolled left, and the right-hand engines lost the airflow they needed. Air France Flight 4590 crashed into a hotel, killing all 109 people on board and 4 more on the ground.

While the human race remains the same genetically as it was many millennia ago, our technological world careens ahead daily. And the pace is picking up. According to Igor Ansoff's *Strategic Management*, the time between invention of new technology and putting it into the field is growing ever shorter. Consider the military. In developing the F-22 fighter, the U.S. Air Force has been surprised to find that the airplane is becoming obsolete before entering service, because key suppli-

FIGURE 1: AIR FRANCE CONCORDE

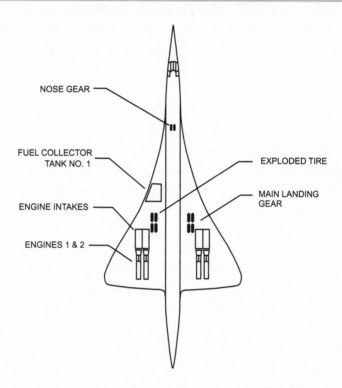

NOSE GEAR

FUEL COLLECTOR
TANK NO. 1

EXPLODED TIRE

ENGINE INTAKES

MAIN LANDING
GEAR

ENGINES 1 & 2

PROBABLE SEQUENCE

1. Main tire on left side hits metal debris on runway.

2. Tire bursts into fragments.

3. Large fragment hits Fuel Collector Tank No. 1.

4. Shock waves from impact burst wall of tank.

5. Leaking fuel pours into left-hand engine intakes.

6. Left-hand engines lose power during climb-out.

7. Aircraft rolls left and crashes.

Adapted from Concorde histories and accident reports

ers have stopped making critical parts. For four decades now our nuclear strategy has been built around deterring attack by intercontinental ballistic missile from one or two enemy states, but this simple plan is going away. We'll soon be facing a half dozen or more adversaries with ICBMs, making false warnings more likely. The U.S. Air Force is developing hypersonic missiles, and the arms race being what it is, the possibility of hypersonic attacks on us will not be far behind. This will give us much less time to contemplate our responses, time that helped greatly in 1979 and again in 1980, when glitches in the nation's computerized missile-attack warning system gave false warnings of a massive Soviet attack.

Overall, the machines in our man-made world generate and deliver electricity reliably, watch through the night for enemy missiles, route cell-phone calls through an ever-changing maze of wires and antennas, and carry people through cloud layers at five hundred very safe miles an hour. Technologic advances appear so easy and comfortable that we forget what Jean-François Pilâtre de Rozier saw right away. This chemistry professor was, by most accounts, the first man to board a hot-air balloon. His opportunity came in 1783 after the Montgolfier brothers of Paris built a silk-and-paper hot-air balloon big enough for a person to ascend in, but couldn't find a pilot. Before de Rozier appeared, the only people to volunteer were two condemned criminals. "I offer my life," said de Rozier grandly, and he went on to make successful flights over Paris.

A lot of us are offering our lives these days to machines and their operators, about which we know very little except that occasionally things go shockingly wrong. And matters aren't much clearer to people on the scene of an unnatural disaster. Bizarre threats appear seemingly out of nowhere, strike, then vanish. But later study shows that machine disasters nearly always require multiple failures and mistakes to reach fruition. One mishap, a single cause, is hardly ever enough to constitute a disaster. A disaster occurs through a combination of poor maintenance, bad communication, and shortcuts. Slowly the strain builds.

Experts found at least six separate mistakes in unraveling the Chernobyl reactor explosion. So it's likely in the future there will be opportunities for people to break the chain and stop a system failure before it reaches a peak. But their chances to respond may be difficult to recognize, fleeting in time, and even more difficult to act upon. Managers and supervisors may resist; workers may not care or know enough to do anything.

I call these events "system fractures" because systems fail in a step-by-step way that's analogous to how metal cracks under stress. Consider a hefty sheet of a flexible metal like aluminum. Though all metal sheets leave the factory with some cracks at the microscopic level, few cracks ever grow long enough to cause the sheet to split apart. But it's dangerous for someone who depends on the sheet, say for an aircraft skin, to assume that none of the cracks ever will grow, because corrosion and gradual fatigue can make them longer day by day. If it reaches a critical point, a crack spreads in "brittle" fashion, at supersonic speed. The whole piece breaks in two with a sound like a gunshot. Slow cracks followed by catastrophic failure happened with two early Comet 1 airliners and more recently with an Aloha Airlines 737 when a length of its upper fuselage blew out aft of the flight deck.

Systems are something like that. That is, all systems of any worth experience human errors and mechanical malfunctions daily. This is not a failure in itself, because much redundancy is built into any mature system. Rather, a failure begins when one weak point begins linking up with others. Even at this stage the failure will proceed no further if some force such as an alert employee intervenes to stop the chain of events, or if the potential failure does not encounter additional weaknesses to help it to fruition.

One premise of this book is that in our new world, surrounded by machines occasionally gone savage, we need to acknowledge the extraordinary damage that ordinary mistakes can now cause and to recognize that as a result a higher level of caution is not only required but is also rapidly reaching down to the level of homes and small businesses. For example, chains of disaster could start with a company that washes airliners, or with a handyman who is working near a gas line in an apartment building. For years dozens of business-minded books have been offering prescriptions for success. Under the business-leadership heading in *Books in*

Print, I found 119 books with "success" in the title. How many books advertised "failure" with equal prominence? One. I propose a revised ratio: for every twenty books on the pursuit of success, we need a book on how things fly into tiny pieces despite enormous effort and the very highest ideals.

INTO THE SYSTEM

You may have toured a large cave with good echoes. If you hooted in the silence, the call would echo, then fade and die away completely within a few more seconds. That's negative feedback, meaning that waves of disturbance die out. Now imagine going into a very special cavern, a sort of mammoth system cave with a positive feedback problem. If you yelled in such a cave, each echo would get stronger and stronger. Eventually you would hear nothing but a continuous roar until the ceiling fell in. That's analogous to what a big system can do at its worst, taking one problem and magnifying it as it bounces off other weak points.

A system capable of going out of control over a span of two thousand miles may sound like science fiction, but it can happen with power grids. Power grids are the most complex systems on earth—so vast that solar flares affect them and so convoluted that the start-up of one generator changes the rotor speed of another generator on the far side of the grid. No person can understand all that a big power grid is doing at any given moment, and sometimes computers can't keep up, either.

Still, it's possible for us amateurs to see the broad sweep of human creations. My background is twenty years of writing on technology and history, with particular attention to why things go wrong. I'm not sure how I acquired this bug. Until his retirement my father was a bulldozer dealer and told us kids many colorful stories about success and failure in the construction industry. To this day, the sound and smell of a turbo-charged diesel engine makes me nostalgic. Like many Americans my life has been touched by machine destructiveness, mostly from motor vehicles: my paternal grandparents were killed in a car crash before I was born; a friend from high school died in a truck crash; a mother in our Boy Scout troop was run down while her car was stopped along the highway, out of gas; and a friend at church lost her husband in the crash of a Galaxy Airlines Lockheed Electra at Reno, Nevada, in 1985, when severe buffeting caused by an open access door on a wing so alarmed the pilots that they crashed the airplane while attempting an emergency return to the airport.

I've always been interested in the survival techniques people have worked out through hard experience, and how these change with new lessons. Extinguishing oil well blowouts once relied heavily on dynamite to deprive the jet of oxygen; firefighters have learned since that less spectacular means like nitrogen and diversion pipes (called "smokestacks" in the trade) work better in starving a flame of oxygen.

It's a privilege to get access to these work sites. After hearing how explosives manufacturing was one of the first enterprises to come to grips with danger, I visited a dynamite plant to see how workers dealt with the risks of nitroglycerine. And hearing that power line workers had developed successful techniques of working on high-voltage transmission lines from a hovering helicopter, I went up with a crew doing that work to see their safety discipline in action. I rode on a cable to the top of a five-hundred-foot broadcast tower so I could get the "tower hand's" point of view. So bold beforehand, I found that standing on a metal framework with nothing but sky above and cows below was extremely unnerving. My steel-working escort had to persuade me to pry my hands loose from the structure and lean back in the safety harness, which is the only way to get any work done up there. And I saw a little bit of hell when Colorado firefighters dressed me up in bunker gear to go in with a crew learning how to subdue a burning propane tank.

Such field trips have given me great respect for the people who work on the edge. Visitors come and go but these men and women keep up such work for years, long after the thrill goes away. They show that what might appear as ultrahazardous labor to the rest of us can be done with considerable safety by those who take the risks seriously and have the right equipment. But in fact all of us face some risk from today's structures and machines. Sometimes it hits right where we live.

In July 1998 the collapse of the top half of a seven-hundred-foot-tall scaffold and mast on the Four Times Square construction job in New York City sent debris crashing through the roof of a residential hotel across the street, killing an eighty-five-year-old woman in her top-floor apartment. And the consequences of even a small system failure can be serious enough for those caught in the culminating moment, which could be a fire, explosion, or car crash.

Homes and offices can have small but troublesome systems. Anybody who has tried to upgrade operating-systems software on a home computer that has peripheral devices attached knows that even small systems can be baffling and sneaky.

Once the average person understands that system fractures usually grow out of a chain of errors and mishaps over time (rather than striking like a bolt out of the blue), that human error falls into certain broad categories, and that people can take action during the early stages to break the chain of causation, he or she should be better equipped to see trouble brewing.

TERRORS AND WONDERS

The world of large-scale industrial and technological disasters is a no-nonsense one, capable of causing chaos and destruction fully rivaling a war zone. Precious seconds roll by as survivors struggle to make sense of the new world around them, comparable to the way a computer needs time to reboot after it crashes. These events can be so traumatic and strange that they lie far from the traumas that trigger our primordial "fight or flight" instincts. People panic when they shouldn't; they sit when they should be running for their lives. One passenger on a crash-landed British airliner recalled that his fellows were so stunned with the impact and with the realization they were still alive, that all of them sat placidly and watched a burning stream of fuel trickle down the aisle.

But if high-power machinery is the domain of strange and terrible events, it is also a place of miracles and wonders. This world gives us outrageously improbable chains of circumstance and stories of sudden salvation against all odds.

Writers of classic literature need heroes and villains, and if we look hard enough we can always find someone holding at least one link in the chain of circumstance. We live at a peculiar time in history wherein a few people without any criminal intent can set loose events that can kill or maim thousands of people. Here the real villain is a chain of circumstances and has no face or name. "Awfully nice people—the nicest people you could hope to meet—cause terrible accidents," says attorney Jim Schwebel, whose firm represents some of the injured families from a crash during a 1998 holiday parade in downtown Minneapolis.

System fractures are not Shakespearean tragedies or Greek plays in which the victims are doomed at the outset by fate. Destiny has no place here, nor is disaster inevitable. Most of us have heard the aphorism, usually called "Murphy's Law," that whatever can go wrong will go wrong, and at the worst possible time. By the end of this book you'll probably agree that it can't be true, given the vast number of near-miss incidents over the years.

What Captain Edward Murphy actually said at Edwards Air Force Base in 1949 was a little different and has real truth in it. Complaining about some instrumentation that a technician had installed incorrectly before a rocket-sled test, Murphy said that if there was any way for this man to do the job wrong, he'd do it that way.

Occasionally things do go wrong at the worst possible moment, such as during the 1981 collapse of two walkways suspended in the lobby of the Kansas City Hyatt Regency Hotel, in which the timing was not coincidental. The crowd in the lobby that night totaled at least fifteen hundred. The same factor that made the damage so horrendous in Kansas City—dozens of people standing on the walkways during a social event—provided the stress that caused the connections holding up the walkways to tear through the steel frames, dropping tons of steel and concrete onto more people below.

Since avoided disasters are rarely newsworthy, newspaper readers and TV watchers may get the idea that disasters are somehow inevitable. But we will see many cases in which people saw the danger and stopped the system's fracture early enough. Take the narrow escape of the dredge *Essex* in October 1989, while it was engaged in clearing the Delaware River channel. Nudged about by two tugboats, the *Essex* was attached with thick wire ropes to a steel pipeline that carried the dredged mud to shore. On October 16, the passing tanker *Kometik* lost control after a power failure in one of its tugboats. The tanker veered toward the *Essex*, on course to capsize the dredge and overrun its crew. The *Essex*'s crew couldn't get out of the way because wire ropes locked the vessel to the dredging pipeline. But the crew was ready even so. As part of a premobilization safety check, Norfolk Dredging employee Bill Murphy realized that the day might come when the cables to the pipeline must be cut in great haste, and had requisitioned a hydraulically powered cable cutter for just that purpose. The dredge supervisor had ordered the machine bought immediately, so the crew had it handy and was able to break free of the pipeline in time. There are many more cases of "salvation," though most are known only to insiders.

Near misses are in this book though they aren't catastrophes. Similarly the book includes mishaps with no injury at all, like the misshapen primary mirror of the Hubble Space Telescope. These hold lessons, too. Although such failures didn't kill anyone, the problems were truly painful to scientists who had spent years developing the instruments aboard and who had planned to work on the project for years afterward. Most of us will never know career frustration as deep as that.

Some incidents have competing theories of cause. I've gone with the official investigation findings, but afterward summarize the other point of view, usually from the company being blamed in the official report. And sometimes it happens that the first government findings are wrong, as in the explosion of the battleship *Maine,* ascribed to an underwater mine in the port of Havana, when an equally likely cause was spontaneous combustion of coal that ignited gunpowder stored just across a bulkhead. According to the most recent studies, an iceberg scrape might not have sunk the *Titanic* had the metal in the rivets for the hull been of good quality.

Some of the disasters in this book were prominent in the news while the drama unfolded. But a few mishaps that you will read about here have never been described in print before or have gotten little attention. In fact, one of these neglected events was the worst technological failure of all time. It started on the night of August 7, 1975, when a massive typhoon stalled over Henan Province in central China. The deluge—more than a foot of rain per day for three days—first caused a few high-country dams to collapse. One was the "impregnable" 380-foot dam at the Banqiao Reservoir, supposedly good against thousand-year rainstorms. Dams downstream couldn't handle the walls of water. Sixty more dams burst that night in the chain reaction; in the ensuing floods, at least twenty-six thousand people drowned.

Even in the famous cases, the public usually missed hearing the full story because investigative reports and trials took months or years to complete, and often these official findings never made headline news. Details of exactly what went wrong with the valve that stuck open at Three Mile Island Unit 2, dumping coolant out of the reactor for more than two hours, are still hazy twenty-two years later because the liability dispute is stuck in the court system.

I've made no attempt to catalog and describe every technocatastrophe of the industrial age. There are far too many. In particular I looked for mishaps combining human error with mechanical malfunction. Some were near misses or caused some destruction but stopped just short of disaster. The appendix has a list of incidents in this book, organized by date and location.

Because most system fractures have occurred out of the sight of the public and news cameras, or at extreme camera range when captured at all, it's difficult to appreciate the energy that the big ones unleash. Even the famous images of the *Challenger*'s orange fireball and white corkscrew smoke trails couldn't convey the energy of the two boosters and why they deserved such careful attention. The reactor at Chernobyl shot flaming chunks of molten uranium high into the air because the fuel rod channels acted like mortar tubes. As the freighter *Grandcamp* burned at the Texas City docks on an April day in 1947, two light planes circled overhead so the passengers could watch the firefighters work. The exploding shipload of ammonium nitrate fertilizer destroyed both planes, a half mile up. When a faulty electronic signal triggered a thrust reverser on one engine aboard a 767 airliner over Thailand in 1991, the craft flipped into a downward plunge so extreme that it broke the sound barrier and disintegrated before hitting the ground.

LIFE ON THE EDGE

Vast machines running out of control might sound like something beyond human ken, but one way to keep matters in perspective is to think of a frontier. Americans know all about frontiers, having lost our western one more than a hundred years ago. No new geographic frontiers have opened for us since then, but a different kind of frontier is well under way and was opening even as the West was being won. It is the "machine frontier," and it is still unconquered. A machine frontier has, in a virtual sort of way, the same characteristics as a geographic frontier: dangers and rewards, bounded at the edges by unknown territory. So if we wanted to plot a movie, prototype testers would be our explorers; facility operators would be the cowboys; and entrepreneurs eager to dominate the new markets would serve as our cattle barons. While some technological dangers can be charted, other risks of a new technology cannot be known with certainty or in some cases even imagined until much later.

Like the old geographic frontiers, machine frontiers attract certain kinds of people. The ultimate machine frontiersperson accepts risk in return for reward or excitement. She does not like to be instructed by anyone on how to behave and does not like poring over detailed safety manuals. She likes the thrill of the unknown, so a mature system whose every quirk is documented holds no interest. She values advancement, exploration, and excitement. She wants to be prepared but is confident that if a problem appears in a high-energy, high-stakes system she'll be able to work out the details on the spot. And if a spectacular crash or conflagration takes her life, that's a better way to go than nodding off in a nursing home. A hacker is a machine frontiersperson; so is an undersea welder on an offshore oil platform.

So stand by for stories from that frontier. I hope you find them salutary, because the price in lives has been high. At a time when wealthy folks will pay $100,000 for an exciting climb up Everest, there's no way for such people to buy their way into the control room of a giant system on the rampage. It's a good thing: too many people who are caught in the maw of the machine discover that there is no way out.

When an organization develops people who are expert at picking out the subtle signals of real problems from the constant noise of routine difficulties, and when the boss allows them to report and take prompt action, that organization is doing the same thing that chess masters do. A chess master spends more time thinking about the board from his opponent's perspective than he does from his own. The story of the *Ocean Ranger* shows what can happen to those who lean too heavily on routine, so certain are they that unpleasant "what if" scenarios will never combine in a single day to make a calamity. But as Winston Churchill wrote in describing the chain of mishaps that let the German warship *Goeben* escape a British task force in World War I, the terrible "ifs" have a way of accumulating.

1: SHOCK WAVE

HIGH TECH ON THE HIGH SEAS

On the morning of February 14, 1982, the crew of the drill rig *Ocean Ranger* began squaring away for a storm predicted to arrive later that day. The men weren't worried; the *Ranger* had been through at least fifty storms in six years of offshore work. It was the biggest floating drill rig in the world when built six years before and was designed to survive hurricane weather of 110-foot waves. Weather reports for the Grand Banks put the coming North Atlantic gale at well below the design limits of 100-knot winds, so no extraordinary measures were called for.

Take a rectangular end table and glue a toy oil well on top; now nail on two additional table legs on two of the longer sides, making a total of eight legs; now glue the legs on top of little pontoons and float the table on a lake with the pontoons sitting just underwater and the rest of the table rising above the surface (see Figure 2). That was the general layout of

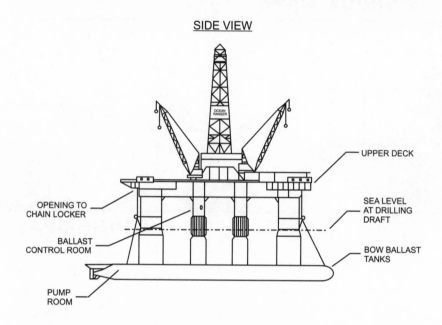

SIDE VIEW

PROBABLE SEQUENCE

1. Wave breaks out "portlight" window to ballast control room, dousing electronics with seawater.

2. Short-circuited valves open and close in ballast tanks.

3. Crew shuts off power, later restores power.

4. Opened valves allow seawater ballast to shift to bow.

5. Crew attempts to empty bow tanks with suction pumps, but rig goes further off balance.

6. Rig's list allows storm waves to reach openings into chain lockers in bow.

Adapted from Royal Commission report

the gunmetal-gray *Ocean Ranger*. The flat top of the table, the double-decked "drilling platform" on board the rig, was home and workplace for eighty-four men. The drilling crew worked in alternating twelve-hour shifts that changed at noon and midnight. They worked seven days a week for two to three weeks, then rotated to shore via helicopter for a paid leave.

Held in position by three strong anchor cables at each corner, the fourteen-thousand-ton *Ranger* took the waves so easily that on a calm day a visitor might be fooled into thinking that its corner legs ran all the way down to rest on the seafloor, 260 feet below, instead of riding on pontoons. Most of the time the decks showed a slow, shallow roll that was barely enough to affect games on the rec-room pool tables. A rig superintendent once called it unsinkable.

Viewed from over the steel railings, the waters below didn't look like much of a menace to men living so high above the water. But the appearance was deceptive, and the men of the *Ocean Ranger* had fresh proof of that. Compared to the twenty-seven years of the offshore oil industry, major rig mishaps had been hitting historic highs in recent years, with twenty-two rigs reporting fires, blowouts, capsizes, or sinkings in 1980 alone. The reason, said industry sources, was the worldwide acceleration in drilling activity. Although shipyards could build rigs quickly enough, finding expert crews to man them was a major difficulty.

In that troubled year of 1980, the *Alexander Keilland*, a floating dormitory for North Sea oil workers, had rolled over in a winter storm, killing 123 men. It happened after a cracked strut in the steel framework finally broke all the way through under the constant pounding of waves. A shipyard painter had painted over it during construction (known because some of the crack length contained paint) but apparently had not reported it. The strut's failure allowed rough seas to break off one of the *Keilland*'s five legs, throwing it out of balance. The *Keilland* took twenty minutes to roll over, spilling the men into the frigid sea or trapping them below.

The *Ranger* had fifty-four Newfoundlanders aboard, a nod to Canadian regulations that encouraged local hires. Two of them were

Stephen and Robert Winsor, brothers aged eighteen and twenty-three. A mishap on board the *Ranger* the previous summer had crushed Stephen's hand, but he went back to the *Ranger* as soon as he recovered. Though the brothers had started as roustabouts without any oil field experience, the rig operator offered pay much better than any land job open to them. An older brother, Gordon, worked on board the *SEDCO 706*, a smaller floating rig stationed nine miles east. The *Ocean Ranger*, the *SEDCO 706*, and a third rig, the *Zapata Ugland*, were all within sight of each other.

According to the schedule made out for February 14, evening was Gordon's time to rest, but the rolling of the *SEDCO* rig in the rising storm kept him awake in his bunk. In almost six years on offshore rigs, it was the worst North Atlantic storm he'd ever experienced. At 7:00 P.M. he got up to walk around. At about that time the rig jolted heavily with the impact of an oversize wave striking the steel legs. Two more big waves followed immediately.

Outside, seawater was "shipping green" over the upper deck, meaning it was so deep that no foam showed. Water even flowed down ventilator shafts into the engine room, so the wave crests must have topped eighty feet. Pumps started dumping ballast water out of the pontoons and the *SEDCO 706*'s drilling platform slowly rose another fifteen feet above sea level. This extra height would make the drilling platform less vulnerable to wave damage; already the waves had bent thick steel I-beams.

Gordon joined his shift on the derrick at midnight. Soon he heard through friends that the *Ranger*, where his brothers were working, was transmitting distress calls over the Telex machine. He hustled to his rig's radar room. Sometime around 3:00 A.M., as Gordon watched, the blip that was the *Ranger* disappeared from the *SEDCO 706*'s radar screen. He asked for sedatives.

When the first reports went out that the state-of-the-art *Ranger* had overturned and sunk in relatively shallow waters, maritime experts were baffled. One theory was that it had succumbed to a strange sort of undersea blowout called a "gas-loaded ocean," where the seafloor would erupt with a geyser of gas that so foamed the water under a rig that the vessel would lose buoyancy and sink. Another theory was that a rogue wave had taken the *Ranger* down, the kind of killer wave that arises out of nowhere, towering ninety feet high or more, the sort that had bashed in the bridge of the *Queen Mary* in 1942, seven hundred miles off the coast of Scotland. Rogue waves are a known hazard off the east coast of South Africa.

In fact, one of the waves that had rocked Gordon Winsor seven hours earlier had done in the *Ranger*, though it was no rogue wave. Divers would find that the wave, by breaking a simple sheet of glass and throwing hundreds of pounds of seawater into the ballast room, had tripped off a long sequence of malfunctions and errors that killed the entire crew.

The *Ranger* disaster lays a good foundation for the subject of clashes between man and machine because it shows how failure ripples through a system like a slow-motion shock wave. Consider how a person can lose his or her emotional equilibrium for a full day or more after an argument or a minor car crash. Similarly, the crew of the *Ranger* lost its poise that night and never regained it.

While following the stormy saga of the *Ocean Ranger* and other conflicts between machines and men, we will look for lessons beyond the obvious errors pointed out by the official boards of inquiry. Observe the fixation on daily routines at these sites right up until the culminating event, without any serious planning to survive an emergency that was well within the bounds of foreseeability.

STATE OF THE ART

The number 2 Eba shipyard of Mitsubishi Heavy Industries finished the *Ranger* in 1976 according to plans drawn up by New Orleans–based Ocean Drilling and Exploration Company (ODECO). ODECO was a seasoned offshore operator. Its owner, Murphy Oil Company, had built one of the first transportable, submersible platform rigs, called *Mr. Charlie*, back in 1953.

Since 1976, the *Ranger* had drilled off the coasts of Alaska, New

Jersey, and Spain before parking in Canada's offshore Hibernia Field, where Mobil Oil of Canada leased it to put down test holes. Three floating rigs worked the Hibernia Field at the time. This was all exploratory work and very expensive, but in 1982 the oil companies and the Newfoundland government hoped that mighty shiploads of oil would start flowing out of Hibernia within five years.

Registration papers filed with the Coast Guard listed the *Ocean Ranger* as a "mobile offshore drilling unit." A MODU like the *Ranger* was paradoxically both more stable and less stable than your typical ship. It was more stable because its deep draft, long legs, and multiple anchors allowed it to ride above big waves more smoothly than any ship, which pitches, heaves, and rolls in heavy weather. But the *Ocean Ranger* was less stable than a ship if it went out of balance, because it was top-heavy. The *Ranger* was 151 feet high from decks to keel, not counting the derrick. Including the derrick, more of the rig's structure was above the water than below it. Furthermore, shifting heavy loads around on the platform could throw the ship seriously out of balance.

Begin from the bottom up. The *Ranger* sat on two giant submarine-size pontoons, each 400 feet long and set 240 feet apart, like catamaran hulls. Inside each pontoon, steel walls divided the space into tanks that held ballast, fuel, and fresh water. When the rig was cruising from place to place the pontoons rode up on the surface, the rig moving under the power of diesel-electric motors and thrusters at the stern. The *Ranger* could and did cross oceans on its own power. When the rig arrived on a station, the crew opened sea-chest valves and let seawater into pontoon tanks as ballast so the pontoons would submerge to about eighty feet below the surface. This made the rig less top-heavy. Waves had little effect when the *Ranger* sat so deep in the water, doing its drilling work. This depth was called the "drilling draft," and it put the decks fifty feet over the ocean's surface, enough for high waves to pass underneath. If a big storm came up, with higher waves possible, the crew pumped out enough ballast to lighten the rig and bring the superstructure another ten feet out of the water. Called the "survival draft," this configuration gave the platform more clearance against damage from giant waves.

Welded on top of the pontoons were eight massive steel legs, rising vertically to support the drilling platform, like the piers of a bridge. The platform consisted of two horizontal decks. The platform was home to the crew and the base of all drilling operations. For an emergency evacuation, the platform held three working lifeboats, ten life rafts, and a helipad.

The support legs were so roomy inside that the designers also used them for storage and working compartments. This economical use of space would contribute to the *Ranger's* demise in several ways: it put a critical electronic control room within wave-splashing range of the ocean, and it made holes in the legs that could allow them to flood in a storm.

The legs placed at the corners of the platform were massive, each almost forty feet in diameter. Each corner leg housed a tall compartment called a chain locker, which had enough space to store thousands of feet of wire rope and anchor chain when the rig was moving around and not secured to the seafloor. At the top of each corner leg, seventy feet above sea level, were three big holes for the wire ropes and chains to come aboard. This was a serious but unrecognized problem. Each hole into the chain locker was a gaping five feet across. Nobody who designed the *Ranger* thought to provide a method of shutting those openings when the rig was on station, nor did they think it necessary to provide some kind of warning to the crew if the chain lockers began to fill with seawater through the holes. Flooding of the chain lockers through the holes wouldn't happen unless the rig was drastically tilted in bad weather, and apparently everyone in charge thought this impossible.

Although the pontoons' machinery compartments could be reached via ladder ways from the surface, normally no one worked down in those dark spaces. One of the smaller, middle legs on the starboard side held the ballast control room. Seated comfortably in a round, metal-walled control room twenty-seven feet above the water's surface, the ballast operator could fill or empty all the ballast tanks by remote control. Dials and lights on a display board told him what was happening down in the pontoons.

On this rig, ballast operators also needed to see the painted draft marks on the legs and what was happening with supply boats outside, so the control room had four round glass windows, called "port lights," set into the steel walls of the leg. These port lights, each eighteen inches across, could not be opened, but they were only made of glass and could break under stress. Part of the ballast operator's duties involved pushing red and green buttons that operated electric solenoid switches. The solenoids controlled a flow of compressed air that ran in copper pipes, down the legs, to the seawater valves in the pontoons. Powered by compressed air, those big seawater valves actually did the work of connecting the pumps in the stern, through pipes, to the tanks spaced along the pontoons. There were sixteen tanks in each pontoon and only three pumps, so to work the ballast system properly, the operator had to set the valves to link the proper pump with the proper tank. When connected through the valves and turned on, the pumps could empty ballast tanks by sucking seawater out of the tanks and dumping it into the ocean. They could also shift water from one tank to another. These pumps were "powerful and capable," an inquiry board said later. But unbeknownst to the operators they had peculiarities that would emerge in mysterious fashion when things began going wrong.

Though the rig seemed stable with its pontoons submerged and all twelve anchor cables rigged out to the seafloor, keeping it level was a balancing act because weight and balance shifted constantly with the busy pace of offshore work and resupply efforts. The *Ranger* could hold almost four thousand tons of supplies and drilling gear, but even that was not enough for weeks of continuous work, so work boats resupplied it every few days. It needed regular deliveries of drilling mud mix, diesel fuel, tools, fresh water, and supplies for the crew. Even between supply runs, the rig could lean off balance as the crew drained tanks and shifted thick steel pipe from the storage rack to the drill string. If a ballast operator failed to make timely corrections to the ballast tanks as new loads came aboard or as the roustabouts worked, the rig's angle would go off center. Drilling work might have to stop, or something worse could happen. Even five degrees off plumb was a serious matter.

One demonstration that the *Ocean Ranger* was dangerously vulnerable to ballasting mistakes came just a week before the disaster. On February 6, as a supply ship pumped fresh water through a hose to the rig, the ballast operator on duty, Bruce Porter, wasn't in the control room. He was down in one of the pontoons working on a mechanical problem. Trying to help out in Porter's absence, the rig's master, Clarence Hauss, sat down at the controls and attempted to correct a slight tilt caused by the water-loading operation. But Hauss unknowingly left one of the inlet valves open to the sea. Tons of unwanted water flooded into two port ballast tanks. The rig leaned almost six degrees off horizontal; it doesn't sound like much, but on the rig it caused much excitement. As the crew hurried to the lifeboat stations, senior ballast operator Donald Rathbun left his bunk and went to the ballast room to investigate. He stopped the flooding and pumped out the two compartments. Later the rig boss, or "toolpush," Kent Thompson, summoned both men to his office for an explanation. Thompson ended the conversation by telling Hauss to keep his hands off the panel and to stay out of the control room entirely unless an operator was there.

While this had to be humiliating to a man who held an unlimited license to command oceangoing vessels—Hauss had served on freighters for fifteen years with Bethlehem Steel Corporation, as master and mate—it showed the split personality of the *Ranger*'s leadership. On those rare days when the *Ranger* was motoring around the ocean to a new drilling location, the master was truly captain of the ship and was supposed to have at least two able seamen for his crew. ODECO brought aboard a "transit master" when it needed to move the rig any long distance.

Whenever the *Ranger* was anchored, the toolpush commanded the vessel and the master did what he could to keep himself busy. It's clear from the context that while the *Ranger* was drilling, Hauss was there only to meet the legal niceties, and not in any real command. Hauss was supposed to serve as backup to the ballast operators in case of problems, but he didn't know the *Ranger* control panel, even though he had been master on two other ODECO rigs before. The late-1970s oil boom had pushed Clarence Hauss back out to sea, from a variety of shore jobs. Before coming to work for ODECO, Hauss had spent ten years working in the jobs of medical technician, stevedoring superintendent, and salesman.

So it seems that the *Ocean Ranger* had left far behind the old days of sail and steam seafaring, when captains truly were the masters of their vessels. These would be people like Captain E. W. Freeman of the tramp steamer *Roddam*. This ship had been caught by the explosion of Mount Pelée in 1902, while anchored in the harbor at Saint-Pierre, Martinique. Half his crew was dead from the blast wave, and the *Roddam* was on fire and trapped by its anchor chain, but Freeman knew enough about the ship to decide he could break free by letting the chain run out as he ran the engines full astern. So it happened that the *Roddam* was able to escape the harbor without help from any deckhands.

The most knowledgeable ballast operator on board as February 14 approached was Don Rathbun, aged thirty-one, who had been doing the job almost two years. Rathbun was a Rhode Islander, born into the Rathbone clan of old New England stock. He was tall and thin; he loved drawing, photography, and spending time out on the ocean. Rathbun had dropped out of college and lived in Rhode Island and California, divorcing and then remarrying. Along the way he owned a carpet-cleaning business and worked as a lobsterman and in a wire factory. Nothing paid as well as the offer he got from ODECO in January 1980: thirty thousand dollars per year beginning pay, no experience required. Rhode Islanders got a rare look at the rig when it moored in Narragansett Bay in August 1979, on the way to drilling in Baltimore Canyon off the U.S. East Coast.

As with many of the *Ocean Ranger*'s crew, the work wasn't something Rathbun talked about much during his three-week shore leaves in Rhode Island. The rig was the rig and the shore was the shore. When he was home with his wife, Nancy, he needed all the time he had to catch up. From reports by men who had worked on the rig, and recollections from family members of the men who died, the *Ranger* was not much different from other rigs. There was some tension between the original American crew members and the Newfoundlanders who were taking their jobs to satisfy Canadian employment requirements.

Some of the men called the ship the *Ocean Danger*. Gordon Winsor recalled that it had trouble keeping men on board. Any offshore rig has its risks, though, particularly for divers or men handling the massive drilling gear. According to the statistics, during the year before its demise, the *Ranger* had about the same injury rate as other rigs. It did have the reputation of drilling during just about any storm, rather than disconnecting its drill pipe from the seafloor. In all its years on the stormy northern seas, the *Ranger* had "hung off" only once. Commonly the men said that the two scariest things about the work were the helicopter trips over open ocean and the possibility of a blowout and fire. The food was good, but the seven-day week, twelve-hour workday schedule left little time or energy for recreation.

The *Ranger* had its little group of practical jokers; one trick directed at newcomers was to jam a man's cabin door shut while he was sleeping and splash water under his door. Then the jokers took up station out in the hall, banging on the door and yelling that the rig was sinking.

One year to the day before the *Ranger*'s final crisis, Rathbun organized a Valentine's Day party for his mother, who was in declining health. He called his parents often via the rig's ship-to-shore radio, checking in on her condition. By this time Rathbun had moved up from roustabout to ballast control operator, which is mostly inside work compared to that of the roustabout he had started as. There was no formal test to take or government-issued license nor was there even a training manual to study.

On the *Ranger*, people qualified for the ballast-control job by spending several hours each day of their sparse free time hanging about the control room and watching over the ballast operator's shoulder. ODECO had once offered a few days of training for new operators but that ended before Rathbun, Domenic Dyke (Rathbun's assistant), and Hauss arrived. So when the operator and toolpush felt comfortable with the newcomer's judgment, after weeks or months, the apprentice left his original job behind and took up tasks in and around the ballast control room. While

ODECO's written training policy called for employees to have eighty weeks of offshore experience before starting their on-the-job training, it wasn't working out that way on the *Ranger* by the time of the disaster. Dyke had started his training after forty weeks offshore, and Rathbun after only twelve weeks. The men who would be called on to save an eighty-four-man crew and a $100 million rig got woefully little preparation, found the investigating commission afterward: "The training program did not provide an understanding of the electrical and mechanical operations of the ballast control system nor the effects of ballast gravitation. A thorough knowledge and understanding of what might go wrong and how to detect and remedy the situation were also lacking. The training emphasis was based on the erroneous assumption that the ballast system was fail-safe."

The *Ranger* did have an operations manual, which was stashed in the ballast control room. It described routine operations of the ballast system and laid out the simple but tedious calculations needed to use the tanks to trim the rig. Operators like Rathbun were told to read it during their apprenticeship.

At some point while whiling away his twelve-hour shifts, Rathbun found out about an undocumented arrangement that would forge a fatal link in the disaster to come. It was intended as an emergency plan and had been requested by ODECO's electrician while the rig was under construction back at Mitsubishi. It was a set of brass rods, stored out of sight inside the console for emergency use only. The hope was that if electric power somehow failed and valves needed to be opened down in the pontoons by remote control before electrical repairs could take place, the operator would take out a set of threaded brass rods. By threading a rod into a corresponding socket on the panel and screwing it down far enough, the operator could bypass the panel's dead electrical system and force a given solenoid switch down. This would send compressed air down to the pontoons and open a specific tank valve in the pontoons. We know that Rathbun had learned about it because he had mentioned the brass rods to another ballast operator a few weeks before. Rathbun had told him that it was an emergency method for closing the tank valves. He was wrong.

VALENTINE'S DAY

Time to take stock, just before the February 14 storm and its ninety-knot winds. The crew was working 184 miles from land in the North Atlantic, two hours from evacuation by helicopter under ideal conditions. The *Ranger* had three and possibly four working lifeboats, but how to use them under the bad conditions brewing was a mystery. Nobody had ever trained for evacuation during a bad storm, or even tried it. Crewmen testified later that abandon-ship drills were jovial events held early on Sunday afternoons. Sometimes the men went as far as lowering the lifeboats on their ropes, releasing the fastenings, and motoring around with the small engines provided. Sometimes they did no more than count heads.

These expensive lifeboats would have been wonderful to have when the *Titanic* sank. The boats were completely enclosed, had a radio and emergency supplies, and could stay afloat in a very bad storm once safely launched. That was the key: the boats had to get from a high deck down to the water sixty feet below, without breaking open. It would have been relatively easy if leaving a ship, the hull of which makes a windbreak on the lee side. But the open steel structure of a rig offers no shelter from the giant waves, and is only something for lifeboats to smash against. The crew's apparent confidence in what these boats could do for the men under storm conditions was completely unfounded.

The rig had plenty of life vests but no full-immersion exposure suits for men floating in a winter sea. Exposure suits are thick rubber-insulated affairs that envelop the body and can keep a man afloat and alive for hours in near-freezing water until rescue. They cost $450 each at the time. Because helicopter rescue could not come fast enough during a storm to save people from being thrown into the cold water, and because the men had no suits to survive immersion in that water, staying alive would come down to a single plan: they had to get the four lifeboats safely down from the deck to the ocean with all the men inside.

Back in Rhode Island, Rathbun's older brother, Robert, saw the snow falling on Valentine's Day, heard about the storm building off

Newfoundland, and thought of his brother Donnie. Robert knew how bad it could be out there; he had seen this area of the Grand Banks firsthand while serving as a radioman on the Coast Guard cutter *Escanaba*. The weather in this area off Greenland, known as Ocean Station Bravo, produced waves so big they cracked steel plates on the cutter and bent its guardrail stanchions flat.

The *Ocean Ranger's* crew did not ignore the storm making up that day. Throughout Valentine's Day the men worked to secure the rig for the blow. As darkness fell, the waves grew to over fifty feet high. About 8:00 P.M. one of them broke out the small porthole window in the ballast control room. Later investigation showed the glass as too thin to meet the original specifications, but even a full-thickness window might have failed under that hammer. As it was, the wave punched the glass completely out, shearing it from the brass frame like a battering ram.

The *SEDCO 706* reported three giant waves at this time, so it could have been one of these that broke out the glass. In hindsight the crewmen should have shut the steel storm covers over the glass before the weather turned bad, but now it was too late to do anything but close them off against more water and try to dry the room out. That still left many gallons of seawater inside the ballast control panel. Salt water is a very good electrical conductor, and once it seeped down into the panel Rathbun and Dyke had no quick way to dry the panel out.

Shortly afterward the radioman aboard the *Seaforth Highlander*, a support vessel standing off to assist the *Ranger* seven miles downwind, overheard a walkie-talkie report from the rig that a window had broken in the ballast control room and men were sweeping up glass. The report also said that anybody who touched the panel was getting an electrical shock. About an hour later the tone of the message was more urgent: from the look of the mimic board, all the ballast control valves on the port side were opening and closing by themselves. Rathbun and Dyke would have taken such an eerie development very seriously, since the misbehaving valves could connect tanks with each other, thus allowing water into the wrong tanks. An open sea chest valve would be especially dangerous because it would allow more seawater into the ballast tanks.

The problem was serious at this point but not fatal. Inspection of the control board afterward showed that some lights had been shorting out and some valves had been opening and closing from seawater short circuits in the microswitches. Because turning off power to the panel made all the valves in the pontoons close immediately, regardless of any short circuits, someone should have shut the power off as soon as possible. But it took time because the correct switch was concealed inside the console cabinet. The rig's electrician arrived, found the cutoff, and doused the power at about 9:00 P.M. Shutting off power also killed the lights and dials on the panel that showed valve positions, pump status, and water depths in the maze of ballast tanks. The only usable instrument without power was a bubble indicator, like a carpenter's level, that showed the rig's lean.

Outside, waves climbed to fifty-five feet, winds gusts hit eighty miles per hour and the air temperature dropped below 0°F. It sounds bad, but had these conditions prevailed, the rig would have survived the night. The ballast tank valves were all shut, and the twelve anchors were holding. But sometime shortly after midnight, the men in the ballast control room made a fatal decision to restore electric power to the ballast control panel. Why isn't known, but they might well have thought they had dried out enough electrical gear to operate it safely. They might have wanted to check out the system, or maybe they wanted to pump out enough water from the ballast tanks to raise the *Ranger* to its "survival draft," where it would tolerate storm waves better.

Because things started to deteriorate from this point, it's likely that their experiments with restored power, combined with persistent electrical short circuits in the control panel, let a surge of water into the bow tanks. A combination of a few open valves would have done it. They might have tried this more than once, shutting the power off between attempts. Because valves once opened took a half minute to close, each attempt would have put more water into the wrong place.

It would have been evident from the inclinometer in the control room that the rig was settling down ever so slowly at the port bow, leaning into the northwest wind. Still, Kent Thompson, the toolpush, didn't request the rig's support vessel to come closer for evacuation. On top of all the other mistakes, that meant that if anything went wrong, nobody would be on hand to pull the *Ranger's* men from the water before hypothermia killed them. Thompson must have thought that the ballast operators could pump out the bow compartments before the situation worsened.

At some point a ballast operator on the *Ranger*, probably Rathbun, decided to use the little brass rods in an emergency attempt to close valves between the bow ballast tanks and the pump room. He kept the pumps running meanwhile to try and empty seawater from the bow tanks. We know the brass rods were used because a diver investigating afterward found that eighteen of them had been threaded into the panel.

But the set of little brass rods was not a good substitute for a working ballast control panel. The rods could only open ballast tank valves, not close them. And without electricity to the gauges, Rathbun couldn't read the instruments to know what was going on down in the ballast system as he experimented with the rods. He had to assume that the equipment was behaving as he expected it to. It wasn't.

THE SECRET OF THE PUMPS

To save the *Ranger* at this late hour, Rathbun or Dyke or Hauss would have had to know some things about the pumps not revealed in the operations manual. They were suction pumps, relying on atmospheric pressure, and therefore strictly limited in their ability to pull up water that was lying much deeper than the pump room. Adding horsepower didn't matter with equipment like this. Rathbun didn't know the crucial fact that once the *Ranger* started tilting toward the bow and the bow compartments went deep, the pumps back at the stern could not suck seawater from the bow tanks when the bow was so much lower than the stern. So under crisis conditions when the wrong valves were open, and when there was no time left to correct any mistakes, the *Ranger's* pumps would do the opposite of what the ballast control operators intended.

A properly trained person would have known that once the rig tilted dangerously toward the bow, he would have to start pumping out middle tanks, and then he could move forward from there, one tank at a time, as the rig slowly came back upright. Although slow, this method would have stayed within the pumps' operating range. But Rathbun didn't know the technique and couldn't see the trouble the pumps and tanks were having because all the electricity was off to the gauges on his instrument panel.

We can only imagine the rage and fear in the ballast room as midnight passed and the rig leaned further into the northwest wind, defying all attempts to get it under control. Perhaps other crew members stomped down the spiral stairs to see what was wrong and to offer desperate suggestions.

Meanwhile the final blow of the technological shock wave rippling through the *Ocean Ranger* was about to hit. It would take effect in the corner leg on the port bow. As with the problem with the suction pumps, nothing in the manual had warned Rathbun or Dyke of this problem, either. The rig's tilt had reached the point where the bigger waves could reach the top of the leg.

The waves threw tons of water through the chain-locker opening and into the cavernous storage compartment below. The chain-locker openings acted like hatch covers left wide open on a freighter's deck during a storm. This growing threat was invisible to the crew because the *Ranger* had no electronic alarms to warn the crew about this extremely serious development. In the kind of positive feedback loop that sinking ships get into, the chain locker's flooding lowered the port bow even further, and this allowed even smaller waves to heave water in through the opening. The crisis was now self-reinforcing and would play out to the end. At some point the drill pipe tore loose from the racks, thundered down the deck, and plunged into the ocean.

Finally at 1:05 A.M. the *Ranger's* radioman received the order from Thompson to call the *Seaforth Highlander* for assistance. Following standard orders for bad weather, the *Highlander* had been lying about five miles downwind, so it wouldn't run into the *Ranger's* anchor cables while keeping station. The rig's tilt had reached the point of no return, nearly fifteen degrees off vertical. Thompson ordered the men to climb aboard the three working lifeboats and release them from the davits. The rig was tilting but still upright, so there was sufficient time for everybody to get aboard. As the fiberglass boats began the sixty-foot drop to the water, two problems became clear. As the boats descended on their ropes, the northwest wind and giant waves hammered the fiberglass hulls against the steel of the tilting rig, cracking them open.

And once on the water, the boats' occupants had desperate trouble releasing the boats from the ropes connecting them to the rig, because each boat had to be sitting calmly on the water, with all weight off the lines, before the rope release would operate. This was ludicrous in a howling storm, when boats fell and rose with each mountainous wave, meanwhile crashing against the steel legs. Within a half hour only one of the lifeboats was still afloat and upright, and that one was staved in at the bow and taking on water.

Having received the summons and running its engines at full revolutions, the crew of the *Highlander* took an hour to claw five miles upwind. When the *Highlander* pulled up at the rig sometime after 2:00 A.M., wrecked boats and life-vest beacons dotted the water, scooting downwind. The *Highlander* stopped alongside the only occupied lifeboat. It looked like salvation had arrived for the eight men alive inside. But in climbing out and trying to reach up to the high gunwales of the supply boat's aft deck, the survivors tumbled off the lifeboat and into the water. Severe hypothermia reduces the smartest adult's thinking to a childlike level, and apparently the men had forgotten under the conditions that the lifeboats were extremely tippy after the men unbuckled their seatbelts. The *Highlander* was not a rescue craft, and its crew had no gear to drag the men aboard in the seconds available before wind and waves carried them off. The survivors were too immobilized by cold to climb into the life rafts that the *Highlander's* crew dropped to them, and these rapidly blew away.

And so, within arm's length of rescue and warmth—at least one of the men actually touched the hull of the supply ship—not one of the eight men still alive from the *Ranger* made it up and over the *Highlander's* rail. Searchers found only twenty-two bodies out of the eighty-four set adrift.

Canadian and American hearings and wrongful death lawsuits followed, turning up a long list of shortcomings at ODECO and Mobil Oil of Canada, along with oversight problems at Canadian and American agencies. Rigs needed survival suits for everyone on board. They needed a way to get boats off the deck in a bad storm. The ballast control room should have had separately powered panels for monitoring and for valve controls. The men in the ballast room should have shut the watertight storm hatches over the glass windows before the storm. The rig should have had precautions or at least warnings against flooding of the chain lockers.

Ballast operators needed expert training on crisis management, and a manual for emergencies. If Don Rathbun had gotten good instructions, investigators decided, he would have known to kill the electrical power to the panel and leave it off until after the storm. This would have caused the compressed-air system to slam all the ballast valves shut and keep them shut. This simple act would have given the rig an excellent chance to ride out the storm without any further tinkering with valves or pumps. But by attempting to work the system with its gauges and instruments blacked out, and its pumps doing the opposite of what they were thinking, the men on the *Ranger* opened a clear path for the pumps to make things worse.

The aftermath? ODECO stopped naming its rigs with the *Ranger* appellation. A later-model ODECO rig that had been named the *Ocean Ranger II* while under construction at the Sumitomo Heavy Construction shipyard got the new name *Ocean Odyssey* after the disaster. The *Odyssey* had its own reckoning, in September 1988, when a well blew out and engulfed that rig in fire and explosions, killing the radio officer.

Salvagers raised the *Ocean Ranger's* hulk for a tow out to deeper water and deliberate resinking, but it broke loose and sunk on the way. Richard Hylund still keeps Don Rathbun's sea cap in his front hall at home, as if expecting him to knock on the door one day and ask for it back. Did Rathbun contribute to the disaster with well-intentioned efforts? It's likely he did, because he was in the ballast room at the time when fatal mistakes were being made and he was the senior man. But

ODECO had not trained him in emergency operations of the ballast system, and he was doing as well as any of us could have done in that position on such an evil night.

"In all the years it drilled," Hylund says, "*Ocean Ranger* never produced a drop of oil."

"The hubris of it all just strikes me," says Don Rathbun's sister Diane. "We heard it was the largest oceangoing rig, it was unsinkable and on and on. And look what ended it—a stupid little porthole."

The *Ocean Ranger* disaster shows how difficult it is for people to sort problems out from the control room, on the fly, as failure starts to spread through a complex system. The people who were supposed to be in control didn't know at least two important things that night: the strange workings of the ballast system when the rig was tilted, and how a storm might take a small off-balance condition and make it much worse. And the slow and ambiguous response of the ballast controls made managing them very difficult once the rig left the familiar boundaries of civilized machine behavior. The next chapter explores the problem of machines whose critical workings were even more deeply hidden from the operators.

2: BLIND SPOT

BAFFLED AND BEWILDERED INSIDE
THE MASSIVE SYSTEM

On January 8, 1989, a British Midlands 737 was flying shuttle service from London to Belfast. The captain and copilot heard a bang at twenty-nine thousand feet and felt the airframe shaking at a high frequency. These occurrences, in combination with the smell of overheated metal in the fresh air gathered by the engines' compressor stage, made the captain and copilot suspect an engine problem.

The flight crew checked out the engine problem not by eye—the limited visibility prevented anyone in the cockpit from seeing back to the engines on the wing—but according to a manufacturer's checklist and by using engine controls and instruments. When the captain switched off the autopilot and pulled back on the throttle to the right-hand engine, the vibration and noise eased greatly.

The situation suddenly seemed much clearer now. Vibration had eased soon after they reduced power on the right-hand engine, so the problem must be there. The crew radioed to air controllers that the flight needed to land nearby and agreed on the East Midlands airport. There was no need to declare an in-flight emergency, because the 737 was fully able to divert and land with a single good engine. They reduced power on the left-hand engine on the way down and the vibration diminished further.

In fact they had leaped from initial hypothesis straight to a fatal conclusion, because they couldn't easily check reality by looking at the engines. The right-hand engine wasn't in trouble: it had been doing fine until they shut it down two minutes after the first noises. Passengers were puzzled by the captain's announcement that he had shut down the right-hand engine, since some could see the *left*-hand engine throwing off sparks and flame. Mechanically, the problem had started when fan blade number 17 had broken as a result of metal fatigue and had thrown fragments into the rear part of the engine, causing more damage.

Nobody from the passenger cabin came forward or contacted the flight deck. It was the captain speaking, after all, and he was sure that it was the right-hand engine that had the problem. The airplane joined the glide slope to East Midlands, and the vibration lessened further. The airliner descended like this for a quarter-hour, the left-hand engine slowly chewing itself up but still providing just enough power to deceive. Two miles short of the runway, the left-hand engine went to pieces, slinging fan blades over a wide area. Now the fire warning went off on the left side. The captain did his best to restart the right-hand engine, but the airplane was only nine hundred feet off the ground and he ran out of time. The jet hit a field, bounced up, and slammed into an embankment of the M1 motorway near Kegworth. Forty-seven people died.

The investigating board found that even without a fire alarm from the bad engine, one small indicator on the cockpit video-display screens could have set the crew straight, namely, a vibration readout for the left-hand engine. Its reading was at five, meaning maximum vibration. But it was a small indicator and out of the way, mostly for engine-trend monitoring and not able to set off its own alarms.

The Kegworth crash shows how problems can happen in any system with a blind spot. This means a machine whose inner workings are hidden from the operator's view and whose instruments do not show up plainly enough to overcome a powerful human tendency to jump to conclusions when under stress. The pilots had a full fifteen minutes to consult the engine vibration monitor or simply to ask the flight attendants whether they could see any visible signs of engine trouble by looking out the side windows. The engine failure would have been no more than a jotted note in a flight log if the operators had taken the time to "consult reality" rather than jumping to conclusions. Another midair crisis occurring out of view doomed ValuJet Flight 592, in 1996, because the pilots had no smoke alarm or any other way of detecting an oxygen-boosted fire in the cargo compartment below in time to return to the airport.

People can do remarkably well in controlling complex machines whose workings are fully understood and open to view. One example is how crews manage to land airplanes on the decks of aircraft carriers at one-minute intervals or less with very few accidents, especially considering the hazards. When meeting a new system, people need time to know its workings under good conditions and bad. The most dangerous time is when the operators don't know what they don't know.

ATOMS FOR PEACE

The most expensive blind spot in history happened on March 28, 1979, at the power plant called Three Mile Island Unit 2, near Harrisburg, Pennsylvania. Though the plant looked fine on the outside after the crisis was over, on the inside the reactor core was a complete wreck. Half the fuel in the reactor vessel had melted, and most of the rest had tumbled into a heap of rubble on top of that. Still, the thick-walled stainless steel reactor vessel held the fuel from breaking out the bottom, just long enough.

The President's Commission on Three Mile Island concluded that the reactor core came within a half hour of total meltdown. Notwithstanding the media's excitement over the "hydrogen bubble" after the first day, it's clear now that the first three hours at TMI-2 were the most critical. While it's possible that the reactor core could have melted into the ground in the so-called China Syndrome, it's equally likely that fuel melting out the bottom of the reactor vessel would have set off a mighty steam explosion when it hit the pool of water in the bottom of the thick concrete containment building, breaking open the building to release a massive cloud of radioactive steam across the towns and farms of southern Pennsylvania. Though news media excitement peaked a few days later, the worst danger had passed on the same day it started and certainly was over by April 1, when President Jimmy Carter arrived at Harrisburg, Pennsylvania, to reassure the public. According to Walter Mondale, during the visit a woman spoke up after Carter said that the danger was over. "I believe you," she said, "because if there was any danger the vice president would be here instead."

The apparent bumbling in the control room inspired cartoonist Matt Groening to give Homer Simpson the job of reactor operator at the Springfield Nuclear Power Plant. Three Mile Island Unit 2 cost General Public Utilities and the rest of us more than $4 billion, making it the nation's worst industrial disaster in cash terms. It was so big that two decades later, lawyers are still fighting over who should pick up the cost.

Beforehand, the Nuclear Regulatory Commission had declared any such "loss of coolant accident," occurring while the reactor was at full power, as so unlikely as to not require consideration in safety plans. And it was a bizarre case of automated equipment trying to save itself, while operators blocked its moves at every turn for the first two hours and twenty minutes. Was it the case of "stupid error" that a British government official called it afterward?

The operators' actions will make more sense after knowing how key parts of the plant were hidden from view, how many instruments gave false readings, and how operators' early training had given them a mindset to close off the emergency cooling. They saw this last, desperate measure as the only way to keep the reactor pipes from bursting. Studies of human behavior in other disasters have often shown this "cognitive lock" phenomenon, where those on the scene decide on a course of action and hold to it against all contradictory evidence. Cognitive lock is a perfect match with an opaque and complex system. Together they have the power to turn a minor problem into a very dangerous situation.

According to measurements at the site, TMI-2 didn't get far enough down the disaster road to release fatal levels of radiation to the outside. In the lingo of the field, TMI was more of a near miss than a catastrophe. TMI vented about fifteen curies of radiation; the Chernobyl disaster seven years later turned out millions of times more. The number one reason the containment building didn't break open was Brian Mehler, a supervisor who wasn't even supposed to be in the TMI-2 control room that day. In fifteen minutes, he figured out what was wrong.

A malfunctioning device straight out of the steam days called the pilot operated relief valve (PORV) was the feather that broke TMI's back. Steam technology was so important to the reactor's troubles during the first, critical three hours that a riverboat engineer from 1850 would be able to understand most of what happened at TMI-2 very well. He would probably call it a blowdown, though, rather than a meltdown. "Blowing down the boiler," in steam lingo, means to dump the steam out through a valve, either preparatory to shutting the boiler down or as a way to loosen mineral deposits from its inside walls.

As much as he would marvel at the fabulous expense and size of a reactor, the old-timer would be amazed at how little the operators could see or hear of what was going on in the reactor building. The reactor building was only a minute's walk away from the control room, but because it was there to contain any leaks of radioactive steam and water, it was not the kind of place you'd work in daily. Operators had no instrument telling them how much water was in the pipes cooling the reactor, and several key instruments would give them dangerously false readings in the crisis.

By contrast, from his post in the engine room while the engine was running, an operating engineer from the steam era could see and hear and feel pretty much everything, short of crawling into the boiler while the engine was running. And he could go even there after he blew down the boiler and let it cool. Sometimes he didn't wait that long.

When the tramp steamer *Tripoli* lost power and drifted toward the rocks of Gibraltar in the late 1800s, the chief engineer crawled inside the firebox while it was still too hot to touch, relying on coveralls soaked with seawater to keep himself alive just long enough to get out. Staying on a plank to keep himself off the grates, he stood up in the chamber to install a washer. The engineer knew to face the firebox door as he did so, so that if he collapsed from the heat, he might fall on the plank. That way other men outside in the engine room could pull him out before he boiled to death.

They had valves showing whether the boiler had enough water, and later, "sight glass" tubes to get the same information more quickly. By the 1850s they had gauges to judge boiler pressure; before that they had to judge pressure by whether the safety valve opened. Some engineers said they could judge pressure by watching the heaving of the riveted boiler walls, in and out, as the cylinders drew steam. Such engineers, then, would be distinctly nervous about the idea of sitting all day in a control room and running a monstrously big boiler by remote control, operating at a steam pressure twenty times greater than the highest pressure of the steam era boilers without being able to see and hear what the machinery was doing.

But, after all, we're only talking steam. When civilian uses for atomic power first hit the public stage after World War II, making plain old water into steam looked like the crude beginnings, just a short bridge to the new era of plenty. According to legend, Hungarian-born physicist Leo Szilard first conceived of a sustaining chain reaction in 1936 when standing on a European street corner. Szilard obtained a patent and secretly assigned it to the British government, but nobody paid attention to him until the bomb-building Manhattan Project. In the early stages of the project, Enrico Fermi headed a group to assemble a crude reactor under the stands of the University of Chicago football stadium, and he welcomed Szilard's input. The world's first reactor was uranium arranged in a pile, with a larger pile of graphite bricks around it. The purpose of the "atomic pile" was strictly experimental, to see what it would take to scale up a bigger reactor to harness energy from a sustaining nuclear chain reaction.

Fermi's pile worked because uranium atoms have a nucleus of neutrons and protons that is easily broken apart when hit by a neutron from outside. In breaking up, the atoms release showers of more neutrons, which in turn break up more atoms. During the Manhattan Project physicists learned how to regulate the number of loose neutrons so the reaction in the pile would turn out a steady heat but not spew out of control; one of their discoveries was that some materials, such as cadmium, indium, and boron, absorbed neutrons easily and if inserted into the pile could damp down a runaway reaction. A reaction running totally out of control was fine if you wanted a bomb, but not otherwise.

After two bombs ended the war against Japan, most Americans felt better about the atomic age angle. It sounded wonderful, particularly to the writers of books like *The Atomic Revolution* (Robert D. Potter) and *Atomics for the Millions* (Maxwell L. Eidinoff and Hyman Ruchlis). Atomically heated open-air stadiums would be nice. Also, according to Pulitzer Prize–winning science writer David Dietz, cars would run on tiny uranium pills, each pill powerful enough to heat a house through the winter. Factories would transmute base metals to gold so cheaply that pipes used in their operating systems might be made of gold to reduce maintenance costs.

We would certainly need rockets and airplanes fueled by uranium. How about diamonds in trainload quantities, forged out of carbon by the irresistible pressure of an underground blast? Or melting the icecaps with the heat from H-bomb blasts? Officials in four southeastern states tried to talk the federal government into using H-bombs to excavate thirty-nine hilly miles for their Tennessee-Tombigbee barge-canal project. In the process of checking all these ideas out, the Atomic Energy Commission (AEC) exploded underground bombs as far west as Alaska and as far east as Mississippi.

But it was harder than it looked on the magazine covers, and it always cost more than imagined at first. The Air Force's delay in building a nuclear-powered airplane "has been of grave concern to many of our clearest thinking airmen," fretted the editors of *Flying* in 1959. In fact, worried the magazine, the delay suggested pre-Sputnik head-in-the-sand penny-pinching.

The AEC's plans for blasting its way across the Western Hemisphere with strings of H-bombs died with the Test Ban Treaty of 1963. And who can say whether science-fiction editor Hugo Gernsback was ever serious about his idea of winter clothes engineered like wearable electric blankets, powered by personal reactors small enough to fit in a pocket? The closest some citizens ever got to seeing the age of wonders, firsthand, was the fluoroscope installed in some shoe stores in the 1950s, for the benefit of customers dying to see a full-motion X ray of how their new shoes fit when they wriggled their toes.

Atomic power plants jumped out of the wish list and into production because the military, needing them for propulsion on board long-range ships, worked out the key details first. The first commercial nuclear power plant, at Shippingport, Pennsylvania, began running in 1957 and was built around a submarine reactor. Things were far enough along now, ten years after the first burst of enthusiasm over nuclear power, to inform the public that some risk was involved. Physicist Edward Teller, while a booster of nuclear power's benefits, had warned an August 1955 international conference in a scientific paper that reactors would always pose some risk of catastrophe, including turning the site of the plant into a no-man's land. And plant officials played up the importance of safety when reporters visited the sixty-megawatt Shippingport plant while it was under construction. Most of it was underground for safety; "Every piece of equipment down to the last nut and bolt is scrutinized with fantastic care," said *Life* magazine. This was important, the article said, because the gear would be sealed out of sight and out of the reach of repair.

The Shippingport power reactor opened for business in 1957. Though very high operating costs dogged that plant from its first day, the nuclear power plant industry grew at an astonishing pace over the next decade, spurred by fuel subsidies from the government and below-cost pricing by reactor builders. Builders and buyers agreed: make the plants big, because bigger would be cheaper per kilowatt-hour. By 1966, the country had fifteen nuclear power plants running, nine more under construction, and twenty-two on order. Some of those orders were for 1,100-megawatt plants, almost twenty times the size of Shippingport.

One of the late-coming enthusiasts was a holding company called General Public Utilities (GPU), which owned three big utilities, including Metropolitan Edison in Pennsylvania. GPU started its push by requesting permission to build a single nuclear power plant on Three Mile Island, the name referring to the length of the gravelly bar in the wide Susquehanna River, ten miles down from the state capital of Harrisburg. Unit 1 was completed in 1974. Then GPU wanted another plant.

Unit 2, completed in March 1978 and the one headed for infamy, had two cooling towers, a building for fuel storage, a two-hundred-foot-high containment dome to confine leaks from the reactor vessel inside, and two large rectangular buildings for turbines and auxiliary equipment. The cooling towers were one very visible sign that, well before the 1979 crisis, the trends in favor of nuclear power just ten years before were shifting fast. *Sports Illustrated* published an article in 1969 claiming that nuclear power plants were killing off fish by dumping their waste heat in rivers and bays; after the controversy caught on, power companies found themselves building cooling towers at extra expense, which dumped heat into the air instead. Combined with public protests against nuclear power that led to more requirements for safety equipment, every new plant after the first wave saw its costs go way up. TMI-2's final cost went from the original estimate of $130 million to more than $700 million. This had to cause real pain at General Public Utilities. According to some who have studied the origins of the meltdown there, part of the explanation for its mechanical problems may lie in the owner's rush to get the plant into electricity production by December 31, 1978, so it could qualify for tax credits.

FEEL THE HEAT

Just as journalism-school students learn to follow the money, it helps to follow the heat in understanding the layout of TMI-2 (see Figure 3). We can think of TMI as three sets of pipes linked together like loops of a chain, working together to move heat from the reactor core to the outside world, taking it through generators along the way that harnessed the heat and turned it into electric power. We'll call the first set of piping the "reactor coolant" pipes. The reactor coolant pipes took high temperature, high pressure water to extract heat from the one hundred tons of hot uranium in the reactor core. The reactor coolant pipes were entirely inside the containment dome, out of view. Traveling inside these pipes, highly pressurized water extracted heat from the reactor pressure vessel and carried it off to a heat exchanger. Inside the heat exchanger, what we'll call the "steam-making pipes" accepted the handoff of heat and carried it from the containment building. Inside the steam-making pipes, ultrapure water flashed into steam. Arriving at the turbine building, the steam passed through a set of turbines and generators, making 880 megawatts of electricity to sell.

Then a third set of pipes, which we'll call the "external cooling" pipes, conducted the waste heat away from the steam-making pipes and carried it outside the turbine building to the top of giant concrete cooling towers. As the hot water trickled down the inside of the towers, some of it evaporated, and in doing so it cooled the water that remained; pumps sent the cooled water back through the external cooling pipes, back into the turbine building to remove more heat. Of the three piping loops, the external cooling pipes were the most visible and accessible to people working at Three Mile Island; even nearby residents could see when that part of the system was operating to bring heat out of the plant, because of the plumes of condensed water vapor that the cooling towers gave off.

The operators and maintenance people were less in sync with the steam-making pipes, since those pipes carried high-pressure steam and very hot water and were not accessible where they passed through the containment building. But the maintenance people could see and even touch this set of pipes where they passed through the turbine building. And by closing isolation valves they could even separate out some of the steam-making pipes from the rest of the system and work on them while the reactor was operating.

Both these loops of piping were visible to the human eye and

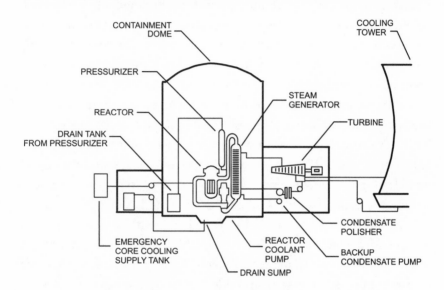

FIGURE 3: THREE MILE ISLAND UNIT 2

PROBABLE SEQUENCE

1. Malfunction during maintenance work causes condensate valves to close.

2. Reactor coolant heats up and pressure rises; pilot operated relief valve (PORV) on pressurizer opens.

3. PORV does not close on automatic command.

4. Coolant begins draining out of reactor through pressurizer.

5. Operators cut back on emergency cooling system and drain more coolant, thinking reactor coolant level is dangerously high.

6. Problem is discovered after half of reactor core melts.

Adapted from Senate Committee on Environment and Public Works open to the wrench, then, in their own ways. Not so the third, the reactor coolant pipes, which held the hottest, most radioactive water. These were entirely out of sight from operators and maintenance workers, buried deep inside the containment building. These pipes and vessels could not be worked on while the reactor was operating, since the pressure, radioactivity, and heat would have made the work unsafe. Not even cameras revealed what the reactor coolant pipes would be doing in an emergency.

Yet all three links of this chain were equally important. The analogy works because a chain is only as strong as its weakest link; at TMI-2, all three loops had to work if the reactor was to stay within its temperature limits. If any of the three loops of piping failed, the reactor would go into an emergency condition, relying solely on short-term supplies of cooling water. The emergency could happen even after a reactor shutdown, after control rods had stopped the critical mass of full-power fission. For hours after a shutdown, the one-hundred-ton reactor core kept producing so much decay heat that the thermal energy could not safely remain inside the pressure vessel without some kind of cooling.

In the early hours of March 28, 1979, the situation on Three Mile Island was as follows. The Unit 1 power plant was down for repairs and testing; Unit 2 was running at 7 million horsepower, enough to supply a city of five hundred thousand people. Big systems like this always have at least a few things out of whack, or maybe more than a few. Down in the hot and cavernous basement of the turbine building at Unit 2, Don Miller and Harold Farst had opened up a part of the steam-making loop for maintenance. It may seem to you like an odd time to do maintenance, but the system had valves to isolate this area from the high pressures and temperatures of the rest of the steam-making pipes.

The problem concerned one of the giant water filters (called a "condensate polisher") in the steam-making loop. All night the men had been trying to shake loose tons of tiny plastic beads in one of the demineralizer tanks. The beads, with the consistency of coarse sand, had jammed into a pipe leading out of filter tank number 7. The bead-jamming problem had plagued the operators on and off for months;

before it had always yielded to blasts of compressed air through a pipe stuck in from underneath. Miller had been trying this for hours, to no avail. The problem was not critical; there were seven other filtering tanks to handle the five thousand gallons of water needed every minute at the steam generator. The reactor was running at 97 percent full power.

At some point during the frustrating hours of fooling around with the compressed air jet and the filter tank, a few ounces of water seeped backward into the compressed air lines, undetected by those on the scene. It took a little while for the water to snake its way into the instruments that relied on compressed air, but then things happened quickly. At 36 seconds past 4:00 A.M., the leaking water from the polisher repair attempt reached the control line to the big valves controlling all the condensate polishers. The automatic controls interpreted this tiny bit of water in their air lines as a deviation from proper conditions and so shut all the valves that let coolant through. This acted like an instant roadblock in the steam-making pipes. The inertia of five thousand gallons of water a minute, stopping so quickly, tore one of the big pipes loose in the turbine building, pulled out controls, and sprayed the place with scalding water. Without water, pumps downstream cut off, so steam stopped flowing from the heat exchanger in the containment building; that made the generating turbines shut down, too.

In 1893, Rudyard Kipling published a poem called "M'Andrew's Hymn," which called for a poet to sing the "Song o' Steam." Three Mile Island sang its own song: every month or so, over the single year from its first operation, escape pipes had howled like an enormous whistle as valves inside the turbine building opened to dump excess steam. That morning, the reactor played its last song as the turbines shut down. For at least a mile in every direction, anyone awake in the Middletown, Pennsylvania, area could hear the sound of a million pounds of high-pressure steam, shrieking into the sky. It woke up one woman a quarter mile away.

So it happened that the three-link chain of cooling pipes suffered a broken link. With the steam-making pipes shut off, the water in the reactor coolant loop had no place to dump its heat. Following automatic "scram" commands from a computer, cadmium control rods plummeted into the reactor, ending the nuclear chain reaction and cutting back heat production at the core to a few percent of full power. Still the temperature rose and the water began expanding, as things will do when heated. The expanding water made the pressure climb from the usual value of 2,150 pounds per square inch.

The reactor coolant loop had only one open space for the water to expand into. It was called the pressurizer tank, and it was forty-two feet high, located inside the containment building along with the rest of the coolant piping. The pressurizer tank acted like a shock absorber to the reactor coolant piping. On an ordinary day the tank was supposed to be about half filled with reactor coolant water at the bottom, leaving steam as a cushion in the upper part. Automatic controls maintained the right balance of water and steam by either cooling or heating the contents of the pressurizer tank.

At the very top of the pressurizer was a safety valve to let off steam if pressure rose too fast for the automatic controls. Deep inside the containment building, the pilot operated relief valve (PORV) opened as intended, reducing pressure by letting steam out the top of the pressurizer when the water level rose. That sent a mix of water and steam down a drainpipe to a storage tank on the floor of the containment building.

THE RUBBLE MAKER

There were many problems at Three Mile Island Unit 2 that morning, but the PORV was the one that reduced the reactor core to slag and rubble. When the pressure stabilized a few seconds later and the electronic command came to close the PORV, the valve stuck open instead. That left a hole about the diameter of a Ping-Pong ball in the reactor coolant system. It wasn't discovered until more than two hours later.

That a safety valve could open and prevent a steam explosion but still somehow cause a terrible problem would have sorely surprised the valve's inventor. His name was Denis Papin, born in France in 1647 and educated as a physician. Papin opened a doctor's practice in Paris but devoted his spare time to his real love, physics. In 1675, Papin moved to England to develop the science of pressurized air. Four years later Papin invented the pressure cooker, initiating a line of thinking that contributed much to successful high-pressure steam engines in the early 1800s.

Papin worked on many projects throughout his life, from air guns to steam-powered pumps to grenades, but arguably the pressure cooker was his gift to the ages. He referred to it often when writing letters to the wealthy, asking them for financial support. He called it the "New Digester or Engine for Softening Bones." Papin recognized that if food cooked reasonably well in boiling water, it would cook even better if the temperature was higher. That means he had to understand steam and how to control it. The ancients knew about steam. According to good authority, Hero of Alexandria directed steam from a teapotlike device and got it to spin a little pinwheel.

Papin understood that the way to make water cook things better is not to build a bigger fire under the kettle. Adding heat only makes the water boil more briskly, without raising the boiling temperature, which at sea level is 212°F. Instead Papin closed off the kettle. That way the water would hit a higher temperature before boiling. Papin fitted a stout U-shaped metal frame over his kettle and equipped it with a screw clamp that pressed the kettle lid down tight. It looked something like an old-fashioned printing press.

It doesn't take much brainpower to realize that a clamped-down pot of water with a fire under it is going to blow up sooner rather than later, so Papin invented the safety valve. Had historians named it the Papin valve people might have remembered him better. Because his most striking inventions remained on paper, Papin never made it into today's pantheon of scientific greats. As it was, he died in London around 1712, penniless and mostly forgotten except in his hometown of Blois, which erected a statue.

To make his safety valve Papin drilled a hole at the top of the New Digester, and arranged a metal weight to sit on the hole to prevent steam from leaking out. To situate the weight he rested a long metal bar on top of the hole, one end of the bar hinging on the cooker. The other end stuck out over the side and was free to swing up and down. Having arranged a lever, Papin hung a metal weight on the free end. By changing the poundage of the weight or by sliding it along the bar, Papin changed the leverage bearing down on the hole of the cooker. This arrangement allowed him to set the pressure that would build before the steam power was strong enough to hoist up the metal bar and vent the kettle. People had used a lever-and-weight arrangement for other devices—it's similar to a balance scale, after all—but Papin was the first ever to use it with steam valves. Today's home cookers are functionally similar to his, operating at about fifteen pounds per square inch with the water temperature at 250°F.

Industrial-size steam valves use springs or motors now instead of weights and beams, but the principle is much the same. The maker of the safety valve that caused trouble at TMI-2 was Dresser Industries, under its Electromatic trade name. The pilot operated relief valve was an electrically operated device, weighing 175 pounds and costing about thirty thousand dollars. At its heart it had a sliding metal cylinder that normally covered a hole into the pressurizer. A spring held the cylinder in place until an electric command slid the cylinder and uncovered the hole, letting steam out of the pressurizer. Babcock & Wilcox used PORVs on its nuclear reactors as a way to help them run more smoothly, to dump steam before the pressure got high enough to open two other safety valves, which were required by law.

Come inside the beige-paneled control room at TMI-2, eight seconds after the valves slammed shut and blocked off the steam-making pipes. The intense investigation into TMI gave us the best record, ever, of how things look to the people trapped inside the fast-moving, scary world of a major system failure. Like Alice inside the rabbit hole, the operators did their best to impose order onto seeming chaos. And they might have succeeded had they been able to see what was really happening inside the containment building only a few dozen yards away.

TMI-2 had four Nuclear Regulatory Commission–licensed reactor operators on the 11:00 P.M. to 7:00 A.M. shift, plus extra men to work on machinery maintenance. The licensed men in the control room at 4:00

A.M. were shift supervisor Bill Zewe and two operators, Craig Faust and Ed Frederick. The shift foreman, Fred Schiemann, was still down in the basement of the turbine building with Don Miller and his partner, trying to fix the condensate polishers.

A two-tone control room warning horn went off shortly after 4:00 A.M., and Faust saw lights warning him that pumps in the steam-making pipes had shut down. Zewe came out of his glass-walled office to join the others at the panel. It showed that three emergency pumps were coming on-line automatically, to keep water moving along in the steam-making pipes and thus assist the flow of heat out of the reactor.

Pressure was going up in the reactor coolant loop; that was to be expected at this point, because of all the heat trapped by the blockage in the steam-making pipes. A bright red glowing light showed that an electric signal had ordered the PORV to open and vent any pressure over 2,255 pounds per square inch. Shortly afterward the pressure dropped and the PORV light went out. Operators thought this meant that the PORV had closed, but all it really meant was that the command had been sent to close it. Nothing showed them that the valve had actually stuck open. The false PORV light was the first deceptive control reading that morning, but not the last.

During the first minute, the level of water showing in the pressurizer tank was dropping, too; also to be expected as the reactor heat dropped and the water cooled. Working from his books of emergency procedures, Frederick turned on high-pressure pumps to throw some extra water into the reactor coolant pipes. This would replace some of the volume lost as the coolant water shrank. The pressurizer water level began to creep upward from its low point of 158 inches. That gauge showed the water level compared to the length of the pressurizer tank, which was a vertical tube. The highest possible reading on the gauge was four hundred inches. Schiemann arrived, having run up eight flights of stairs from the turbine building basement. Schiemann took a seat at a panel on the left side of the console, where he could watch the water level in the pressurizer.

The rising water should have stopped but didn't. When it hit three hundred inches at four minutes into the crisis, Frederick cut back on the water injection pumps, but the water level still went up even though pressure was dropping. TMI–2 was now cruising somewhere outside the operators' familiar world. Past four hundred inches on the gauge, the operators knew, there would be nowhere for rising water volume to go.

Think of filling a closed set of metal pipes with water, right to the top, leaving no air space at all, then sealing them up tight. Reactor operators call that condition "going solid." If you heated those filled pipes just a little more, the increasing water volume would create so much pressure that the pipe would have to burst. In a reactor the coolant water would turn to steam, and the voids left in the pipes would let the core overheat. It would be the much-feared "loss of coolant accident." Said the TMI operating manual in unusually clear words for a technical manual, the pressurizer "must not be filled with coolant to solid conditions (400 inches) at any time except as required for system hydrostatic tests."

Based on what the operators knew at the time, and given the readings from the pressurizer's water gauge, their fear made sense. Training on the Babcock & Wilcox computerized simulator at Lynchburg, Virginia, had never shown this behavior, and in fact the training had stayed away from showing the operators any complex, multiple-failure scenarios like the real one the four men faced here, a scenario that required them to fly blind. In all this big control room, there was not a single instrument or gauge that told them how much water actually sat in the reactor vessel, or in the convoluted piping with its high points and low points, or in the steam generator. Only the pressurizer tank, with its relief valves sitting at the top, had a water level indicator. It was a gauge with a needle showing against a vertical scale, and it was labeled "PZR Level." It lay just inches from the pressure gauge for the pressurizer.

Meanwhile, undetected, the PORV on the pressurizer was blowing down the boiler, letting steam out at the rate of 220 gallons of water per minute. It was this leak at the top of the pressurizer that was sending the water level so high, fooling the operators. There were other people, though, outside the plant, who had already found out how the water level in the pressurizer could fool operators, given how hard it was to see inside the containment building. To these people the apparently contradictory behavior (pressure dropping and water level rising) would not have seemed so strange. Well before March 1979, word had come to Babcock & Wilcox, and the Nuclear Regulatory Commission, as well, of a near miss at the Davis-Besse Unit 1 reactor near Toledo, Ohio. On September 24, 1977, an electrical glitch there had opened and closed a PORV so many times that it stuck open; this caused the water level in the pressurizer to go up as the pressure went down.

Just as would happen at TMI, the Davis-Besse reactor operator had followed his training and shut off the emergency cooling out of fear that the reactor coolant piping was about to go solid. Why no disaster at Davis-Besse? That plant had been running at low power, and the operator had seen his error after just twenty minutes. The Davis-Besse plant changed its in-house procedures after this to warn against shutting off the emergency cooling. But on the reasoning that this brand of relief valve was unique to Davis-Besse and thus of no interest to other power plants, Babcock & Wilcox and the NRC sent out no warnings to operators of other B&W reactors about the erratic reactor behavior that a stuck-open relief valve would cause. Nor did the manufacturer change its manuals or reactor training regimen to warn operators during training that if a relief valve ever stuck open, it could mislead them into thinking the reactor coolant was going solid when it was not.

So, without the benefit of this lesson from Ohio, Zewe and his operators guarded vigilantly against any tendency of TMI to go solid. As the water level continued to go up, they cut back the flow of high-pressure cooling water from five hundred gallons a minute to a bare twenty-five gallons a minute. Feel free to criticize their decision to shut off the high-pressure cooling, but also know that the operators did not design the control room that gave them so many problems that morning. Ed Frederick had sent a memo a year before to management, saying the control room layout was going to cause a severe problem some day. Nothing substantial was done to fix the problems in the control room's alarm system during the months leading up to the final accident, Frederick testified later, and the meeting he asked for never happened.

And still the water level rose on the pressurizer gauge. Zewe and the others now faced a full-fledged mystery.

"A GIGANTIC AND SENSELESS FEAR"

Although the operators' lives were not in immediate jeopardy, what the reactor was doing was fully as unnerving as other incidents have been along the machine frontier, like the moments leading up to the crash of AeroPeru Flight 603 on October 2, 1996. The 757 airliner took off that night from Lima with its left-side "static port" sensor tubes taped over. Removing the tape was a forgotten task after the plane's preflight washing. None of the mechanics or pilots had detected this extremely serious problem before flight. It was serious because those few inches of tape blocked in an erratic fashion the air supply needed by critical flight instruments, particularly the airspeed indicator and altimeter. As recorded on the cockpit tape recording, the AeroPeru crew struggled mightily to make sense out of the lies their instruments were telling them. They were experiencing what sociologist Karl Weick calls *vu jade* (the opposite of *déjà vu*), the profoundly frightening impression that the world no longer makes sense and that one has blundered into a place or circumstance so alien that no one has ever been there before. Weick quotes Freud on this: "a gigantic and senseless fear is set free" when things get bad enough.

The Peruvian airplane swerved, climbed, and descended almost at random as the pilots struggled with the controls and the autopilot. Distracted by constant alarms, the crew members never figured out why the airspeed indicator and altimeter varied wildly though the airplane was flying well. Even with assistance from an air traffic controller, the plane crashed into the Pacific after a desperate half hour's struggle, killing sixty-eight. Its altimeter read 9,700 feet.

The TMI operators stuck with their theory that the reactor was going solid. Maybe a pump or valve somewhere had gotten stuck and was forcing more water in, defying all commands. If cold water was coming in, that might explain why the pressure was dropping. It was hard to concentrate, with the main alarm Klaxon going and more than a hundred alarm lights flashing. So the operators began letting water out of the primary coolant piping at the rate of 160 gallons a minute.

Finally, having cut off the inflow of nearly all emergency cooling water, and having opened a valve to let more water out, at 4:06 A.M. the operators saw the water level stop climbing as it reached the top of the gauge. It looked like a hairsbreadth escape from the do-not-exceed point of four hundred inches. For the next two hours, the water level gauge would hold the operators' attention like a cobra in a basket. It would drift down when they let more water out of the primary system; then it would creep back up toward four hundred inches. It was so strange that Zewe sent his assistants to cross-check the other instruments, to see whether the water level gauge was showing bad information. The word came back later that the gauge was correct.

Other workers and managers on the graveyard shift began to gather, consulting in low tones and offering to watch panels under Zewe's direction. The control room had only a single phone line to the outside world. Even the computer was far away: although the computer was capable of recording key information about hundreds of alarms as fast as they came in, it could only print out fifteen lines of information per minute. The printer fell more than two hours behind at one point in the emergency.

By 4:20 A.M., the combination of choked-down emergency pumps and coolant water flowing out the letdown valve seemed to be working well enough because the pressurizer water level was hovering somewhere around 370 inches. So Zewe left the control room to see if he could straighten out the mess at the condensate polishers. But when Zewe got back at about 5:00 A.M., affairs had taken a very serious and unexpected turn. The four giant pumps in the containment building that forced water through the reactor coolant pipes were shaking themselves to pieces as the pump impellers met cavities of steam, sped up, then slammed into solid water and slowed down.

Pumps and fittings could not survive this abuse without cracking. Worse, this didn't fit at all with the operators' prevailing theory that the primary system was nearly full of water. Instead it was a strong signal that the water was full of steam, meaning that the water level might be dropping below the top of the reactor core. Following standard procedures once again, at 5:14 A.M. the operators shut off two of the pumps, and the last two a half hour later.

It was about 5:00 A.M., when the four reactor coolant pumps began shaking, that an engineer used the precious single phone line from the control room to call Brian Mehler at his home in Palmyra, Pennsylvania. They needed him as backup and wanted him to come as soon as possible. Mehler was not assigned to Unit 2 at the time; the duty roster for the day showed Mehler coming in at 7:00 A.M. to work as a shift supervisor at Unit 1. Mehler had been working over at Unit 2 the day before, and someone must have thought he might have an opinion as to what was going wrong.

Mehler got in his car and hurried down Pennsylvania 421 to the plant, stopping at the gate to check with the guards and pick up his radiation badge. The guards didn't know anything about the emergency, but Mehler could see that the power generators had stopped because no white plumes rose from the cooling towers. Mehler drove to the control building and walked into the Unit 2 control room shortly after 6:00 A.M. The station manager gave him a short briefing.

It was a wild scene: at least fifty engineers, supervisors, and shift workers had crowded into the control room, all trying to make sense of what the controls were telling them. The panels showed 110 alarm lights flashing. The master alarm siren was still driving people crazy. Free for a few minutes to pursue his own train of thought, Mehler pondered how it could be that water pressure was down if the pressurizer's water level was up. Nothing in the manuals had explained it, so something else must be going on.

MEHLER'S DISCOVERY

Mehler approached the problem a little differently from most of the operators. Zewe, Schiemann, Frederick, and Faust, caught in the center of the storm since it started, were all navy men of at least five years' experience and had operated reactors in submarines or on aircraft carriers. Mehler had served his military time in the air force. The only reactor Mehler had ever operated before TMI Unit 1 had been the little open-pool training reactor behind the hockey arena at Penn State University, doing startups and shutdowns.

Fifteen minutes after arriving at the control room, Mehler had two theories in mind. One explanation could be a blown circuit breaker that had knocked out the electric heaters in the pressurizer tank. These heaters were supposed to come on if pressure dropped, making more steam at the top of the pressurizer. Mehler figured that if the heaters weren't working, the system would have trouble keeping the water level down where it was supposed to be. Mehler sent a man to check on the breaker panel.

Without waiting for an answer about the circuit breakers, Mehler turned to the other possible explanation, a fairly small leak somewhere in the reactor coolant system, and one that had opened very early in the sequence. It wasn't practical for him to run over to the containment building, throw open the door, and look for spurting steam. Any clues would have to come from the same two-thousand-plus instruments and controls that had been baffling the others for more than two hours.

Mehler knew that the air pressure in the containment building was up, and it was getting warm in there, which indicated a steam leak somewhere in the domed building. He went to the computer terminal and called up temperature readings for the drain line coming off the PORV. The computer indicated the drain temperature from the PORV was higher than normal, at 280°F, but far short of the reactor coolant temperature of almost 600°F. Mehler could have stopped here, reasoning that since the line was well below the coolant temperature, he was only seeing evidence of the same old slow leak that had been in progress since October; in fact it had first been written up more than a year before, when the plant was still in testing. All the operators knew about the PORV's slow leak and also that the plant had been unable to fix it.

Actually the drain line was much hotter, but a programmer had instructed the computer not to show any drain line temperature over

280°F. Still, even that struck Mehler as too hot under the circumstances. Could this be the leak? Not according to the men thronging the control room before Mehler arrived. Bill Zewe had already asked for a report on the PORV drain temperature earlier. The man who checked on it mistakenly told Zewe that it was close to the usual temperature. And, Zewe believed, the control panel proved that the PORV had shut. Finally, Zewe had twice asked for information about the drain tank that the PORV emptied into, and twice he was told that it didn't seem to be filling. That had been enough for a very busy and harried supervisor. Zewe had not pursued the issue any further.

Mehler asked the supervisors in charge if they minded if he went ahead and closed an electrically operated valve that would isolate the pressurizer tank from the PORV. The actuating switch for the valve was within reach, right there on the central console alongside the indicator for the PORV. If steam was blowing out the PORV, whether a little or a lot, the "block valve" would choke off the leak. They told him to go ahead; nothing else had worked.

One operator told federal investigators later that Mehler's idea was a last-ditch measure, but it wasn't. Blocking off the pilot-operated relief valve posed no risk because the pressurizer tank had two more safety valves, completely separate from the PORV. The two valves, which operators called the "code safeties," were required by the Boiler Code of the American Society of Mechanical Engineers. By law, operators cannot block off the code safety valves or mess with them in any way. The code safeties were set to open at twenty-five hundred pounds per square inch, well below the measured strength of the piping.

At the time Mehler got his clearance to proceed, the system pressure had dropped to nine hundred pounds per square inch. He leaned over to Fred Schiemann and asked him to close the block valve. Within seconds, a gauge in the control room relayed news about the first good thing to happen on Three Mile Island in the last two hours and eighteen minutes. The primary coolant pressure was heading back up.

Core damage was well under way by then and would continue for much of the day; and it would be eleven hours more before the operators brought the water level up enough to cover the core, but

TMI's worst overheating had been stopped less than an hour from disaster. At least one other man had been sniffing along the trail of subtle clues that led toward the stuck PORV, a Babcock & Wilcox engineer named Lee Rogers who was in touch by phone, but it's impossible to know what would have happened had Mehler not spotted the problem when he did.

"I brought a fresh pair of eyes into the room," Mehler says now, but it took more than that to divine the machine at its worst moments and stop the leak that the others had let go for more than two hours. He had the time and the inclination to take his two theories and test them fully against the facts, rather than stopping halfway as the others had done. He could see no further into the containment dome than anyone else. Mehler was doing what economist Herbert Simon called "satisficing," which means coming up with a workable and fast-acting solution without complete information. In an emergency, satisficing is better than "optimizing," which means trying for a solution that is close to perfect.

The primary system had lost about two-thirds of its coolant, allowing half the core to melt; perhaps twenty tons of uranium had melted into a slag pool at the bottom of the reactor vessel. The first direct sign of how serious the damage was became apparent the next day when a water sample was taken. It was loaded with black grit and radioactive particles from the ruptured fuel rods. The full story of the first day's chaos in the containment building would take fourteen years and a billion dollars to sort out completely.

There's still no agreement among TMI experts about why the PORV stuck open; possibly a buildup of boron compounds from the reactor coolant water, or an electrical malfunction—called chatter—that made the valve cycle open and close so many times that it wore out.

OF NORMAL ACCIDENTS AND HIGH-RELIABILITY ORGANIZATIONS

The bizarre events of TMI-2's man-machine struggle attracted much attention from system-safety thinkers. One of them was Charles

Perrow, who cited the events at Harrisburg to open his landmark book, *Normal Accidents*. Perrow concluded that some classes of technology are inherently open to chains of failure, whether sooner or later. With such machines, adding more safety systems only raises their levels of complexity. He said the systems of most concern are those involved in transforming dangerous substances (such as in a recombinant DNA lab or a nuclear reactor) and that show the characteristics of "tight coupling" and high "interactive complexity."

"Tight coupling" means a time-driven system in which one event leads to another in short order. A down-home example of tight coupling would be a line of fast-moving cars in traffic with too little separation between bumpers. Even a small problem with the front car, such as a flat tire or the driver's inattention, can cause a massive pileup among all the rest. An interactive and complex system is one that is subject to chains of unexpected failure. It resists thorough understanding by its managers and operators, like the reactor at TMI-2. Perrow believes the only sure way to manage the situation is to shelve the technologies that are too complex and too likely to fail catastrophically (he had nuclear power plants and gene splicing labs in mind), and to run any marginal ones with a very disciplined organization.

On the other side of the aisle are several sets of researchers who believe that people can safely handle just about any risky business if they organize themselves into "high reliability organizations," with employees who are empowered and trained in special ways. According to the high-reliability school, these groups all share a few key elements: a priority on safety from top to bottom; deep redundancy so the inevitable errors or malfunctions are caught in time; a structure that allows key decisions at all levels; workers who keep their skills sharp with practice and emergency drills; and a premium on learning lessons from trials and errors. These guidelines arose out of observations at such demanding places as air traffic control centers, nuclear power plants, aircraft carrier landing control centers, and nuclear submarines. These researchers zeroed in on plants whose operations have the worrisome elements listed by Perrow but who nonetheless have impressive safety records.

The feeling that mastery over machines is possible has inspired companies like DuPont, Lever Brothers Worldwide, and the INCO Corporation mining conglomerate to adopt a "zero injury" goal for their employees. DuPont reports that some of its plants have gone more than twenty years without any lost-workday accidents.

Though it's generally true that old and tested systems lie more open to view than new ones, my point is not that all modern systems are opaque and mysterious, whereas all old systems are visible and safe to operate. Certain machines of our time are already close to perfection. Today's gas turbines provide much more reliable propulsion for airliners and helicopters than the piston power plants they replaced, for example.

The problem of the blind spot can show up in any system with misreading instruments, hidden equipment, and a sluggish response to the controls. The problem grows much worse when crews don't know how little they are seeing into the machine, and then in a crisis they erroneously cobble up a theory and stick to it against all evidence. The next chapter looks at another hazardous mindset: the zeal to finish a great project before its time.

WHEN FLAGSHIP PROJECTS
RUN OUT OF TIME

All of us can hope we never experience the waking nightmare that Roger Boisjoly lived through on January 28, 1986. At a teleconference the night before, regarding the launch of the space shuttle *Challenger*, Boisjoly had done all a loyal engineer could do to persuade his company, Morton Thiokol, to persuade NASA not to launch the shuttle during the bitter cold temperatures predicted for the next day at Kennedy Space Center.

At the start of the 8:45 P.M. EST teleconference between offices in Utah, Florida, and Alabama, things looked good for Boisjoly and several others at Thiokol who wanted NASA to hold off on the launch until warmer weather. Solid rocket manager Allan J. McDonald warned NASA managers that as company representative he would not sign the launch recommendation. He advised NASA not to proceed until the weather warmed enough to bring the boosters up to 53°F. Otherwise the rubber "O-ring" seals connecting the booster segments might leak, let gas burn through the steel casing, and cause a catastrophic failure.

NASA representatives replied over the telecon link that the leaking-seal problem was a known quantity and under control, and they added that there was no conclusive link between cold O-rings and leaks, anyway. After all, one of the worst gas leaks had happened when a booster was very warm. NASA engineers felt Thiokol's proposed lower limit of 53°F was not supported by any evidence and was inconsistent with the lower temperatures that Thiokol had already accepted during previous wintertime launch attempts. Booster manager Larry Mulloy wrapped up this point of view by demanding over the telecon line, "My God, Thiokol, when do you want us to launch—next April?"

After a thirty-minute break, Thiokol's engineer managers reported back that they would withdraw their objections. Since McDonald wouldn't put his name on the recommendation, booster-program vice president Joe Kilminster signed a form instead and faxed it to NASA. Boisjoly opened his journal that night and jotted down his feelings of anger and worry. At work the next morning, Boisjoly stopped by his boss's office to say that he hoped for a safe flight but also wished the seal on a joint would leak just enough to prove beyond all doubt that a serious problem existed. After Boisjoly left and was walking down the hall, ignition-systems manager Bob Ebeling asked him to step inside a conference room and watch the launch on the big TV. Boisjoly said no, but Ebeling insisted. Boisjoly worked his way to an open spot in the front of the room, sitting on the floor with his back against Ebeling's legs. Sixty seconds into the flight, Ebeling said a prayer of thanks for a safe flight. The *Challenger* broke up thirteen seconds later, painting puffy white streamers across the sky and throwing the room at the Wasatch Division headquarters into a dazed disbelief.

Boisjoly spent the rest of the day in his office, not even able to speak when people stopped by to ask how he was doing. He was not surprised the next day when a fellow employee told him that a video-tape during launch showed a flame leak through a joint in the booster case, just before the disaster.

According to Diane Vaughan's *Challenger Launch Decision*, the definitive book on the subject, NASA and Morton Thiokol messed up because they conditioned themselves to rationalize and then tolerate a growing problem for the sake of bureaucratic goals. Never bringing the problem directly to the astronauts for their input, never taking the danger seriously enough to stop the program for a proper fix, NASA tried to contain the O-ring crisis by fiddling with small things such as insulating putty and the O-ring testing procedure, not even willing to delay the final launch to wait for warmer weather unless Thiokol came up with something more compelling than its technical judgment. It was the "normalization of deviance," according to Vaughan.

"There is only one driving reason that a potentially dangerous system would be allowed to fly," wrote chief astronaut John W. Young after the disaster: "launch schedule pressure."

DREAMING GREAT DREAMS

The British hydrogen-filled dirigible *R.101* and the space shuttle *Challenger* were both megaprojects born out of great national aspirations. The same high-flying promises it took to get them started forced them to keep going forward, despite specific, written warnings of danger by key technical people. In neither case, as in so many other predisaster intervals, did those well-intentioned, urgent memos make a shred of difference. The grim history of memos goes back at least to 1788, when engineer Thomas Telford warned the elders in charge of the grand, four-hundred-year-old St. Chad Church at Shrewsbury, England, that the masonry and timbers supporting the bell tower were fatally weak. Apparently expecting to hear instead that modest repairs would do, they declined Telford's warnings. The church collapsed early in the morning of July 9, a few weeks later.

To those in charge at NASA, caught between deadlines and the problem of never-enough money, the fact that a booster hadn't burned through so far was proof of safety. Understandably, the public's reaction afterward upon hearing of the warnings was anger and disgust. Apparently they didn't agree much with the ancient philosopher Cassius Longinus, who wrote: "In great attempts, it is glorious even to fail." Citizens wanted to know: Were those working on the project crazy, or just lazy? The answer is neither. The people behind the failures were the same ones behind the successes. They were the best and brightest of the aerospace world, and they had been working themselves to rank exhaustion. Their work hours were horrendously long, their commitment absolute. So how could it happen? Doesn't earnest effort count for something?

Each disaster shows in its own way how projects have a way of plowing forward despite bright red flags of danger. With the *R.101*, we'll see the driving personality of airship booster Christopher Birdwood Thomson. Sometimes such a man drives a project to great heights, sometimes to immolation. But he was accountable; Lord Thomson died aboard the *R.101*, along with forty-seven others. And the *Challenger* shows us how easy it is to pass over little glitches when there are so many bigger things going wrong and there is no time to stop and sort them all out.

The wreck of the airship *R.101* was so much a rehearsal for what happened with the *Challenger*, fifty-six years later, that we will see their stories in parallel.

THE BIG SHIP

Begin with the *R.101*, which at 770 feet long was the world's biggest airship when it crashed. Mrs. Shane Leslie got a good look at the aircraft on its way into history that night. She had been eating dinner at her cottage near Hitchin, England, when she heard the servants yelling. She looked outside to see a "ghastly red and green" light shining across the grounds. She joined the staff out on the grass. A giant dirigible, diesels roaring, was coming straight at her house. In the glow of its navigation lights she jumped over a fence as the servants ran in another direction, all of them convinced the house would be demolished in a fiery wreck. But the airship cleared the trees and the rooftop, too, and by such a tiny margin that Mrs. Leslie could look through the windows and see people in the dining room. She said later that "horror descended on us all" as they watched the taillights dwindle. It was a compelling sign that the airship was so heavy it was running at the ragged edge of its capabilities.

A brief history: Nineteenth-century experiments with teardrop and cigar-shaped airships showed that balloons rose well and could even motor around the sky, but they had no strength to carry big payloads. The Germans' solution was the dirigible: a line of gasbags held inside a rigid metal framework, with a smooth, streamlined skin covering the whole thing. The skeleton was a series of metal or wooden ribs regularly spaced along the length of the ship, held together by backbonelike girders running from stem to stern.

Although World War I proved that a hydrogen-filled gasbag had no place in aerial combat, airships did look well suited for long-haul routes and naval reconnaissance. From 1910 to 1914, Germany's five passenger zeppelins made two thousand trips with no crashes. Big airships could pack tons of cargo, and their flights were comfortable, even elegant, at speeds of up to seventy miles an hour. Early airplanes were rickety, a pain for passengers, and short of range and load. Airplane engines had to run on gasoline, which people of the era thought a frightful fuel.

After initial resistance from the Royal Air Force, boosters of British dirigibles put together enough support from the navy and governors of far-flung possessions like Australia and India for a major airship development program to bring the Empire closer. The plan began in 1923 with a proposal for five privately built airships, but after a change of government to the Conservative Party in 1924, the plan turned into a competition instead. It was a horse race of heroic proportions: private sector versus public sector, each building a single giant airship at least seven hundred feet long capable of flying to Australia, with fueling stops at mooring masts spotted around the commonwealth. Britain would manufacture and sell the winning design to other countries. Went the argument: a private company might skimp on safety, whereas the government enterprise could spend whatever it needed to advance the state of the art.

On one side was the Airship Guarantee Company, a subsidiary of

Vickers Limited, building a ship to be called the *R.100* or, more popu-
larly, the "Capitalist Ship." On the other side was the Air Ministry itself,
working out of the Royal Airship Works at Cardington, Bedfordshire,
building a "Socialist Ship," the *R.101*.

Even before Neil Armstrong put his right foot down on the moon
in July 1969, the Apollo program was shutting down. The budget cuts
started in 1966 and chopped 140,000 jobs out of the aerospace industry
within two years. And after the excitement of the first landing, public
interest fell as steeply, with only a brief surge during *Apollo 13*'s crisis in
space. Was it surprising that later landings drew yawns instead of viewers?
America had promised to put a man on the moon before 1970. The first
flight put two men there months before the deadline, with no apparent
difficulty. America recovered so quickly from its moon fever that just a
week after the *Apollo 11* landing, an opinion poll showed a solid majority
of Americans believing that NASA was getting too much money.

At the time of the *Apollo 11*, a presidential panel headed by Vice
President Spiro T. Agnew was finishing up work on a post-Apollo
plan. The report came in September 1969: Given a reusable space
shuttle and a space station, NASA could launch an atomic-powered
Mars mission before 1986, maybe as soon as 1981. It would need a
guaranteed budget of $8 to 10 billion per year. But one by one the big
plums, such as the Mars Mission, dropped off as the Nixon administration and Congress turned their money to other matters. Then, in
January 1972, with NASA's budget down to a third of the Apollo high
point, President Richard Nixon agreed to develop a space shuttle:
publicly because the craft would make space exploration easier and
cheaper, privately because it would bring many jobs to important
political territory. It would cut the cost of launching a pound of cargo
into orbit by at least a factor of twenty. The shuttle should even turn a
profit. Finally, NASA told the press, it could accomplish its goals
absent any further major technological breakthroughs. Apollo had
done most of the work.

The *R.101* was also capable of great things, on paper. The government promised in 1924 that in two years the great gasbag would be
in the air, and soon afterward carrying one hundred passengers and fifty

crew members on a steady time schedule. Cruising at sixty-five to seventy miles an hour, it would reach Egypt in three days or less, a full two weeks faster than any steamship. It would be a submarine spotter for the navy, a troop transporter for the army, and an aircraft carrier for the RAF. Drawings showed it carrying four airplanes inside a hangar, then dropping each plane as needed. "The versatility that is to be demanded of her borders on the magical," said an editorial in the *Evening Standard*.

The *R.101*'s chief designer was Royal Naval Air Service lieutenant colonel Vincent Richmond, an engineer of enormous energy and dedication. He had learned airship design from the Germans but had never designed an airship himself. Nevertheless, he went straight to work on the biggest airship in the world, mainly because there were so few airship designers. The 1921 crash of the British airship *R.38* had thinned the government ranks in this field. "One of the most fortunate of men," as Richmond called himself, he finally got his chance to join in world-class airship work. Richmond and his wife moved to a cottage outside Bedford so he could put in extra hours without interruption, working sometimes until dawn in his study.

The project made a lot of work for Richmond and his assistants because of the Air Ministry's determination that the *R.101* would be a test bed for breakthrough ideas. For example, the frame was made partially of low-tech steel and partially of aluminum alloy. Normally the ribs of dirigibles were like giant bicycle wheels—light, with spokelike wires for strength. The *R.101*'s designers had a different plan: each rib of the framework would be strong enough to support itself with no strengthening wires. This required much extra design work and extra metal to make sure the ribs would be stiff enough. The *R.101*'s creators followed the principle that when in doubt about a girder, make it stronger—and therefore heavier. It had a factor of safety of four, meaning that each girder was able to carry four times the expected stress.

As costs rose past the original estimate and the date of first flight stretched beyond 1927, questions from Parliament to the Air Ministry during debates grew ever more cynical. "When are you going to bring your two old horses out of the stall?" one member of the House of Commons demanded of the air minister.

PUSH THAT ENVELOPE

By 1972 NASA settled on the basic layout of the space shuttle: two solid rocket boosters to help with acceleration (the first time that solid rockets had been used for manned flight in this country) and rocket engines burning hydrogen and oxygen for the full eight-minute flight to orbit. Because the "orbiter" couldn't store enough fuel and oxidizer to run its engines, these liquids would be carried in a huge cylindrical container called the external tank and would flow across to the orbiter through big pipes. In flight, the two boosters would fall away first, splashing into the ocean under parachutes for recovery and reuse. The external tank would fall off later, on the way to orbit, burning and breaking up on the way down. After finishing its mission, the orbiter would return to Earth by using small self-contained rockets to start reentry, then land like a glider.

As development and tests began in the mid-1970s, two enormously difficult problems concerning the orbiter devoured most of the experts' attention. One was the revolutionary system of thirty-one thousand reusable, protective tiles that the orbiter—the manned and winged shuttle craft—needed on its belly and wing edges to survive the blowtorch heat of reentry.

The other hurdle was the hydrogen-fueled space shuttle main engine (SSME). A shuttle needed three SSMEs, each of which would pack more power in less space than any rocket motor had ever been able to, along with tremendous durability to withstand reuse. The turbopumps on each main engine forced liquid hydrogen and oxygen into the system at a half ton per second. The pumps were about the size of a car's engine block, but each turned out sixty-three thousand horsepower, rotating at more than six hundred times a second. Engine tests began in 1975 and problems continued through 1979, with all major parts failing at one time or another. In 1977 alone, four engines went to pieces. Marshall Space Flight Center of Huntsville, Alabama, supervised development of both the main engines and the solid rocket boosters. The main engines didn't approach reliability until late 1979, less than two years before first flight.

If we think of Marshall as the stern parent and the shuttle's propulsion components as offspring, the main engine was the brilliant but erratic child, hogging all the attention. That left the solid rocket booster in the role of the sturdy and dull brother. It was very hard, even impossible, for those outside the solid rocket booster program to believe that anything could go wrong with such simple devices. Each "solid" held 550 tons of propellant in a long steel tube, sealed on one end, with a nozzle on the other (see Figure 4). The only moving parts in a solid rocket motor during its two minutes of useful thrust would be the machinery needed to point the exhaust nozzle.

Because the manufacturer was in Utah, and because a fully assembled booster was too big to move across land in a single piece, each rocket motor traveled to Kennedy Space Center by railroad, broken into four main cylindrical segments. Workers at Kennedy stacked the segments together vertically and topped the stack with the nose cone, making up a full-length booster. Thiokol called the connection between each segment the "field joint" because the work was done in the field, meaning outside the factory. Each booster needed three field joints along the fuel-containing length. To visualize how a field joint works, imagine trying to string short pipes together into a high-pressure pipeline in such a way that you could disassemble and reassemble the whole pipeline every few months. You couldn't weld the pipes; you'd have to use some kind of mechanical joint with removable fasteners. One way to do this would be to provide the front end of each pipe with a slotted rim that the rear end of the next pipe could slide into. That's basically how the booster segments went together at Kennedy: pipes joined end to end, with each rim-and-slot junction held fast by 177 steel pins. For an ordinary pipeline this would have been fine, but not for the high-powered, hot-firing boosters. One challenge was that at each field joint a thin air gap remained between the solid fuel castings in each segment. Without some precautions, flame would fill this gap during a launch and attack the half-inch-thick steel of the booster's outer casing. To keep the flame at the core of the booster where it belonged, the field joints had heat-resistant putty to close off the gap between the fuel castings, and two rubber O-rings fitted into the rim-and-slot arrangement as a final seal.

FIGURE 4: SPACE SHUTTLE *CHALLENGER*

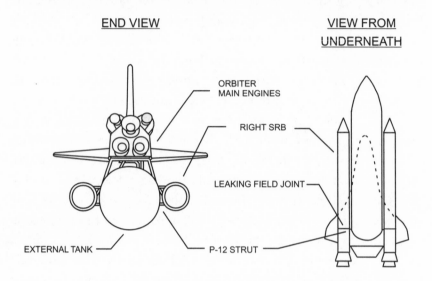

END VIEW

VIEW FROM
UNDERNEATH

ORBITER
MAIN ENGINES

RIGHT SRB

LEAKING FIELD JOINT

EXTERNAL TANK

P-12 STRUT

PROBABLE SEQUENCE

1. Cold temperatures before launch reduce sealing ability of O-rings inside Solid Rocket Booster (SRB) field joints.

2. Exhaust gas leaks from aft field joint of right-hand SRB for first three seconds of flight, then stops.

3. Resuming at 58 seconds into flight, jet of flame weakens rear, P-12 strut between external tank and SRB.

4. Strut fails and tank ruptures as nose of SRB rotates into external tank.

5. Spacecraft breaks up, dumping fuel and throwing orbiter sideways into supersonic slipstream.

6. Crew compartment falls away with fragments.

Adapted from NASA and President's Commission

"Solid rocket, solid technology" was the message of George Hardy, booster manager at Marshall Space Flight Center when he featured the solid rocket boosters at a press conference on October 14, 1980. NASA wanted "to maximize the state of experience on solid booster systems from the past and minimize pushing the state of the art," he said. Doing things this way kept costs down by limiting the test firings needed in Utah. NASA was scaling up the Titan 3 solid rocket motors, even sticking to the same vendors. Because the shuttle motors would be launching people, NASA would increase the margin of safety by adding one more O-ring to each field joint to prevent the leaking of hot exhaust gases, which could burn a hole through the steel case.

The press went baying after Morton Thiokol later, alleging that "inside baseball" was responsible for its booster contract with NASA. Some reports suggested that the company (which was the second-lowest bidder) was chosen because NASA administrator James Fletcher was from Utah and wanted to help old friends.

Whether or not this allegation was true—Fletcher denied it, and an investigation by the General Accounting Office didn't turn up proof of wrongdoing—it wouldn't be fair to conclude that the company was ill-prepared for the task. Thiokol had some of the oldest credentials in the big-booster business. Its motors were vital to some outstandingly successful missile programs during the 1950s and 1960s. Thiokol's founder, Kansas City chemist Joseph Patrick, invented the rubber compound that would make giant solid rocket motors practical. The adhesive even inspired the company name, which was concocted from the Greek words for *sulfur* and *glue,* two substances used by Patrick when he attempted to create a new type of antifreeze and instead stumbled onto a foul-smelling formula for artificial rubber.

A variation of that rubber was in high demand during World War II as a thick, resilient liner for fuel tanks on aircraft. It allowed tanks to withstand gunfire with minimal leakage. After the war, Thiokol president Joseph Crosby noticed that the California Institute of Technology's Jet Propulsion Laboratory was buying many buckets of the company's liquid rubber. He investigated and learned that the JPL found it to be an excellent binder for the oxidizer and fuel powders it was using for solid rocket motors. Thiokol decided to exploit this toe-hold in rocketry and in 1949 began manufacturing seventy-five-pound infantry support rockets at the Redstone Arsenal near Huntsville, Texas. This led to a military contract for the world's first big solid-fuel rocket, the Hermes, weighing five thousand pounds.

After Hermes, Thiokol won the bid to make the sixty-five-thousand-pound first stage for the new Minuteman 1 intercontinental ballistic missile (ICBM). Thiokol moved to a ten-thousand-acre spread in northern Utah and was in the game big time. At its production peak, Thiokol was building two Minuteman first-stage motors per day. Perhaps most relevant for the Shuttle contract, Thiokol worked with the air force in the 1960s to develop supersize solid rocket motors, one of 156 inches in diameter and another monster coded the "260," which would have been twenty-two feet across, intended for heavy lifts into orbit and so big that only barges or ships could have moved it. That idea never flew, but it gave the company the confidence to try for the shuttle booster contract, which it won in 1974.

The fuel Thiokol used on the shuttle solid rocket boosters was almost exactly the same as on the Minuteman ICBM: aluminum powder and rubber binder, mixed with ammonium perchlorate to provide oxygen. This mixture is very stable if left cool and dry. The air force once tested a Minuteman motor after letting it sit in a silo for twenty-nine years. It burned fine.

NASA ordered seven test firings from Thiokol's test stand in Utah before the first launch in 1982. All tests included the effect called "joint rotation," which is the slight bulging of the steel cases upon ignition and which tends to ease open the seal at each field joint if not guarded against. One test even included a trial during cold weather, at an air temperature of 36°F.

One thing NASA did not ask Thiokol to do in the seven tests of an individual booster was reproduce the way the booster would writhe and bend under actual launch conditions. There were at least two dynamic forces that would be acting on the booster, and each worked in concert with low temperatures in January 1986 to bring about a failure. One force was the "twang," a bending of the whole shuttle stack after the main engines lit but before the boosters came on. It caused the whole craft to bend backward about three feet, measured at the nose, and then rebound forward, like a flagpole sways in the wind. This was one force acting to pry open the field joints a small fraction of an inch.

The second force acting to pry the joints open (also not tested in Utah) was the stress accumulating at the struts attaching the boosters to the shuttle's big external fuel tank. Each booster had two lower struts, each mounted to a steel ring around the rocket's circumference. The lower strut numbered P-12 was just a foot from the hole that would appear in the *Challenger's* booster on the day of disaster.

Vincent Richmond and his assistants finished up the plans for the *R.101* in 1927. Steel girders started going up in the vast shed at Cardington that year. The giant dirigible was ready for the air in October 1929. The airship as it emerged from the shed was the biggest in the world, at 720 feet long, bigger than anything the Germans with their decades of zeppelin experience had attempted. Propulsion came from big wooden propellers on five "power cars" hanging from the belly by struts. The *R.101* was so streamlined that the space for crew and passengers lay almost entirely within the great skin.

In its operating style the *R.101* borrowed its traditions from both sea and air. The crew dressed in naval-type uniforms. The airship had rudders and an elevator like an airplane, but the commander in the control cab gave his orders to "coxswains." One coxswain operated the wheel that steered the ship up and down, another handled the wheel for left and right turns. If the pilot wanted to speed up or slow down he used a shiplike engine telegraph to dispatch his orders to engineers in each power car. During the day the navigator steered by compass and landmarks; at night he could go up a ladder to the top of the gas-bag and navigate by the stars. But unlike an oceangoing vessel, this behemoth could not throw down an anchor in harbor or on shallow seas. The *R.101* couldn't refuel or even stop without a giant mooring mast or a giant shed to park in.

Visitors to the Royal Airship Works shed at Cardington came away awed by the size. Less obvious was how the ship was a strange blend of old and new, massive and delicate. Other airships used simple valves that let out excess pressure in the gasbags to keep the skin from bursting when they flew too high. The *R.101* used new, highly sensitive valves that had the problem of dumping gas every time the airship rolled from side to side.

The structure inside the skin was heavier than usual, all girders sized with an extra margin of strength. In this age before plastic film, the sixteen gasbags along the *R.101's* length were made out of many thousands of cow intestines, cut into sections, lapped together, and varnished. The big bags, weighing half a ton each and confined in great steel-mesh nets, worked well enough, but only if kept free of moisture and if kept from rubbing against the metal girders.

Any big project must face one or more crises, and for the *R.101* the first came with its load tests in the hangar. The airship's weight proved appallingly high, and this massive error would feature prominently in the crash a year later. Originally planned to weigh ninety tons without hydrogen or fuel, the *R.101* came in a full twenty-three tons heavier than that.

The most glaring reason for the weight problem was its engines: seventeen tons of Beardmore diesels designed for powering locomotives. Parliament had directed the use of diesels, partially for political reasons and partially in the belief that the heavy diesel fuel was less prone to fire than gasoline. Diesels weighed twice as much per horsepower as gasoline engines and were so heavy that designers had to put extra steel into the dirigible frame to support them. Only later would it occur to the men at the Royal Airship Works that even with all this extra weight the *R.101* was still at risk of a gasoline fire, because each power car had to carry drums of gasoline for the small gasoline-powered engines that cranked the big diesels when starting them. The realization that the *R.101* was flying around with the worst of both worlds came too late to replace all the gasoline-powered starting engines.

Still the men were keenly aware of some risks. Those at the airship works knew that having so much flammable hydrogen around was a real danger. The Air Ministry banned airplanes from approaching within three miles of the ship. Once when the commander of the airship, Major G. H. Scott, saw a man standing under the airship take out a box of matches, he rushed over and knocked him down with a kick. All people going up the mooring mast had to surrender their matches and tobacco, and only one room in the airship was rated safe for smoking. In this room, chains held the ashtrays and lighters to the tables.

In lifting trials the *R.101* could hoist only thirty-five tons instead of the sixty tons planned. But it was enough for making local test flights around England, so with only minor changes the *R.101* began its trials in October and November 1929. On one of the trips it passed sedately over the royal palace at Sandringham, inspiring waves from the king and queen. The trials went smoothly enough, the ship taking to the air only on dry, windless days.

KICK THE TIRES AND LIGHT THE FIRES

On April 12, 1981, astronauts John Young and Robert Crippen took the space shuttle *Columbia* up for its first test flight. Close examination on orbit showed that sixteen of the black tiles on the topside were missing, having been pried loose by the shock of launch. But the critical tiles on the bottom of the shuttle stayed on, so the craft was spared having a hole burned through its aluminum skin upon reentry two days later. And the arrival at Edwards Air Force Base in California proved that the "flying brickyard" could indeed land safely despite having no engine power at the time and thus no second chance.

Successful flight or not, the Reagan administration cut $604 million from the space budget about this time. NASA knew that flying once a week was impossible for the time being. But it still promised routine access to space, and the schedule for the coming years showed a steady push in the pace. By 1984 it was clear from the numbers that things were not going as quickly as planned. That year, NASA had promised twelve launches but delivered five.

NASA came under intense pressure to maintain a reliable launch schedule for its satellite customers. It proved impossible even with all four shuttles flying. Now it was clear what a tough bargain NASA had driven for itself. Before the shuttle, NASA had been able to set its own launch schedule to meet the man-moon-decade goal. Now its customers set the pace, or wanted to.

There were many reasons for the clash between promise and reality, but they came down to a machine that was immensely complex and resisted all attempts at making operations as routine as an airliner flight, which some Reagan administration officials felt was so close they had started discussions with airlines about commercializing the operations. It was far premature. Tiles needed close inspection after each flight, and engines and control systems kept showing distress. In its November 1983 flight (STS-9), the *Columbia* suffered a major electronics breakdown and a fire in an auxiliary power unit during its descent. The APU's hydrazine fuel exploded as the shuttle was parked. The mishap could have destroyed the spacecraft in flight if it had happened a little sooner.

A visit to any one of hundreds of shuttle suppliers would have shown the rank impossibility of a once-per-week schedule. The work was far too exacting. Consider Morton Thiokol's role in particular. Each pair of motors, called a flight set, accumulated hundreds of thousands of documents as it moved from Kennedy to two plants in Utah for final cleaning, filling, and curing of the propellant, and then for return to Kennedy for assembly with the rest of the parts to make a full booster. For example, each expended motor went to Utah with as much as nine tons of rubber insulation glued to the interior walls. All of that scorched and clinging scrap had to be cut and sandblasted off; then each section had to be scanned for defects by the biggest X-ray machine in the world.

The solid rocket boosters were simple but ominously powerful: powerful enough to do 71 percent of the heavy lifting for two minutes, pushing two thousand tons far past the speed of sound. NASA liked to tell the press that each booster turned out as much power as all the engines on seventeen 747 airliners at full thrust.

Greg Katnik, a technician at Kennedy Space Center who supervises the final inspection of the launch complex before each flight, once told me about the fury of the solid rocket boosters. He explained that remote-controlled cameras surround each shuttle before flight, aimed to photograph critical parts of the launch sequence in case of mishap. To protect them from the heat and blast, each camera has a steel cover weighing about eighty pounds. During the *Challenger's* launch in April 1984, the booster exhaust caught one of these camera covers. The force of the blast ripped the cover from its bolts. The cover passed under the *Challenger's* external tank, heading south, then crashed through a six-inch-thick concrete wall and kept going. Completely crumpled, it rolled to a stop four hundred feet away.

The *R.101* floated back into its hangar in late November 1929, sheltering there for the next six months. In December the Air Ministry agreed to major changes, though not to the extent of changing the five power plants from diesel to lightweight gasoline models.

The engineers of the Royal Airship Works thought up three ways to give the *R.101* more lifting power: lighten the ship by pulling out any-thing unnecessary, add enough length to fit another gasbag, and make room for more hydrogen in all the bags by letting out the wire harnesses holding in the gasbags. All the actions had side effects. Obviously, remov-ing sections of inner structure or adding length might weaken the airship. Less obviously, expanding the gasbags increased the opportunities for the bags to chafe against the girders, which would open leaks. When workers removed the bags for inspection, they found that fifteen of the sixteen bags had pinhole leaks too numerous to count. The Royal Airship Works finished two of the changes by April 1929, but for the next three months political problems kept workers from beginning the long process of cut-ting the ship in half and lengthening it.

First the work had to wait so the *R.101* could appear at the annual RAF Air Display to be held in late June. This would help quell public impatience. Soon after the airship left the shed to get ready for the event, a 140-foot rip opened in the silver-painted canvas of the right-hand side. Another great hunk of the cover tore the next day. Riggers climbed up to inspect the damage and came down with the terrible news that the cover was rotting away all over the vast acreage of the dirigible. There wasn't time to replace the cover before the air show, so men with patches, thread, and needles climbed up to seal the rips as the ship floated at anchor at the mast. This quick fix allowed the *R.101* to get off as scheduled on June 27, 1930.

Ralph Booth, captain for the privately built rival airship *R.100*, rode along while the *R.101* prepared for its appearance at the air show the next day. He was alarmed to see the airship drop nine tons of water ballast during the day to maintain altitude, even though it had used up two tons of fuel along the way. It even had to dump fuel on one flight to stay aloft. This was the clearest possible sign that the airship had too little lift.

After the *R.101* reached 102 hours of flight test, only half the hours originally thought necessary before receiving its airworthiness certificate, the flying trials halted once more. England had just one big mooring mast, and the rival *R.100* needed that mast to prepare for its flight to Canada. The *R.101* went back into its shed. This enforced idleness would have been an excellent time to add the extra bay, but that was out of the question because Sir John Higgins, a high official of the Air Ministry, wanted to make sure that one of the two airships took off for Canada on schedule. If a serious problem arose with the *R.100* during preparations, the *R.101* would have to be ready to take her place. Thus, major repairs would have to wait.

It was at this point, July 1930, that a dispassionate observer could have seen the depth of the Air Ministry's impatience with the *R.101*'s progress. Regardless of any public assurances about the importance of safety, the working priority was flights first, safety second. The *R.101* was so short of lift that it needed lengthening; its gasbags were leaking twenty-two thousand cubic feet of hydrogen a day; and its outer envelope was failing in great rips in mild summer weather. Regardless, this airship was supposed to start across the Atlantic on short notice, as is, if the *R.100* faltered.

The situation greatly alarmed at least one man, F. McWade. McWade was the chief inspector at Cardington for the Aeronautical Inspection Directorate, and he held the authority to grant or deny the vital "Permit to Fly" and the airship's certificate of airworthiness. On July 3, McWade defied the rules by sending a confidential memo directly to the Air Ministry office in London, bypassing the usual chain of authority. McWade's memo explained that he had only renewed the airship's Permit to Fly for a few weeks, because urgent fixes were needed. He said that the gas leaks were very serious and the cloth padding around the girders was not solving the problem. Furthermore it risked corrosion to the steel girders because the pads were attracting water. McWade wrote that it would be best for Cardington to pull out every bag, each weighing a half ton, for repairs. McWade ended by saying that under the circumstances he could not support extending the Permit to Fly past July 19.

McWade's memo went first to the chief of his department. The chief knew he was supposed to pass the memo on to the secretary of the Air Ministry, but instead he contacted R. B. Colmore, director of airship development at the Royal Airship Works. It was Colmore's job to both promote the use of airships and oversee their safe flight. Asked for his comments, Colmore wrote back in a reassuring tone that padding around the girders was a good remedy and would be done soon. This was so soothing that McWade's boss never sent the warning letter onward and upward to London. Instead he wrote back to McWade saying he had better make sure the padding was well executed. McWade, his duty to warn completed, dropped the issue. The R.101 got its flight permit extended when the time came.

Finally, more than three weeks after the R.101 went into the hangar to pass the time, Cardington received permission to add the extra bay. In five weeks of nonstop work to meet the flight schedule for Lord Thomson's October meeting in India, riggers cut the airship in half, added 45 feet of length to make the total length 770 feet, and replaced the entire rotting skin.

By 1929, Christopher Birdwood Thomson was the driving force to complete the R.101. Holding the cabinet position of Secretary of State for Air, Lord Thomson was the sort of fellow that technical people have trouble dealing with, but who under the right circumstances could drive subordinates to extraordinary achievement. Thomson favored bold and prompt action, having first gotten attention as a young engineer lieutenant in South Africa when he cleared a blocked railway line promptly and decisively. That maneuver had impressed Lord Kitchener, who became Thomson's mentor. Thomson had served out the war, turning to diplomatic duties afterward. At every point in his life his boldness and verve had served him well.

Lord Thomson believed that airships would dominate air travel for at least the next two decades. The *R.101* would lead the way. In January 1930, during a debate in Parliament, Thomson said: "This is one of the most scientific experiments that man has ever attempted, and there is going to be no question of risk—while I am in charge—being run, or of any lives being sacrificed through lack of foresight."

Regardless of growing misgivings among his subordinates as fall approached, he held no fear for the safety of the "old bus," as he affectionately referred to the *R.101*. "It's as safe as a house," Thomson once told the press, "except for the millionth chance."

THAT STILL, SMALL VOICE

In 1977, four years before the shuttle's first flight, engineers at Marshall Space Flight Center had begun circulating memos about their worry that the field joints on the solid rocket boosters might not hold the high pressures of the first ignition, peaking at almost one thousand pounds per square inch. Each booster had three field joints, which were splices that connected the four segments of each booster upon assembly in Florida. But the engineers' concerns receded into the background during the first ten flights, since few missions showed signs that hot gas had gotten past the insulating putty and reached the first set of hard rubber O-rings that helped stop combustion gas from escaping through the field joints before it reached the exhaust nozzle. But in 1984 something changed for the worse, possibly a test procedure that used pressurized nitrogen gas to check on the field-joint seals. During more than half of the next fourteen shuttle missions, there were signs that hot gas was scorching the first of two O-rings. On some of these flights, gas even touched the second, outer O-ring, meaning the gas had almost reached the outside of the steel case.

The worst damage occurred during the January 1985 cold-weather launch of the *Discovery*, when five O-rings sustained heat damage. You may have seen the effect of cold weather on rubber. A garden hose left out in winter weather turns as stiff as a pipe regardless of whether it has water in it. During the post-accident investigation, physicist Richard Feynman would demonstrate the principle by dipping a segment of an O-ring into a glass of ice water.

In August 1985 top officials from Marshall flew to Utah to talk over the problem with Morton Thiokol management. Roger Boisjoly, a seal engineer with Thiokol, got the assignment to put together a "Seal Task Force" to come up with a quick solution.

After three months of work, the *R.101* emerged from the shed on October 1, 1930, leaving barely enough time for a departure on October 3. The men were close to frantic by Thomson's insistence on taking the airship to India and back in time to attend a conference in London on October 20. From Thomson's view, the Royal Airship Works had already spent two years more than expected to make the airship flightworthy, and now it was time to finish the job, regardless of the setbacks. He and his predecessor as Secretary of State for Air had fought for their higher expenses against stiff opposition in Parliament and had taken much abuse for the delays. The men got the message: the future of the airship business in England depended on getting Lord Thomson to India on time.

With the new gasbag slotted in and the ship lightened up, there was enough time for a single test flight. The *R.101* left its mast on October 1, returning seventeen hours later. Due to engine troubles, the crew did not attempt to see how the ship handled at full speed, and of course nobody knew how the *R.101* would behave in rough weather because none of the test flights had ever been run during a storm. Perhaps the crew could continue flight tests with passengers, on the way to India.

Simple human haste probably played a role in the sinking of the *Titanic*, according to metallurgist Dr. Tim Foecke of the National Institute of Standards and Technology. D. Colvilles & Company, the Scottish company that supplied the bars of wrought iron that went into the *Titanic's* three million rivets, was under pressure at the time to complete other big projects, and they had too few skilled artisans to meet the demands. The iron went out the door with four times the normal percentage of slag inclusions, and there was no good quality control method to counter this problem at the time. This slag made the rivet heads more prone to shear off under the grazing collision that the *Titanic* suffered, opening the hull at plate seams. We know from undersea observations that the ship sustained neither giant rip nor gash; instead the size of all the openings, taken together, was about the square footage of a throw rug.

Thomson wanted to leave on October 3, a Friday, but Colmore persuaded him to delay to give the crew some rest because they had been working nonstop since the test flight. They agreed that the *R. 101* would leave for the conference on October 4. In that way it would still arrive on time.

Told days before the flight by the director of Civil Aviation that the ship was not ready, Thomson shot back that anyone who was afraid didn't have to go. Proof of his confidence in the ship was the great pile of trunks and suitcases he brought along, estimated by the mast's elevator attendant as equivalent in weight to twenty-four people, though the flight crew's personal baggage was under strict control. The ship also carried rolls of carpet and extra provisions for a formal dinner planned in Egypt, along with extra diesel fuel so the airship wouldn't have to refuel during the party; the smell might spoil the food's aroma.

Likewise, predictions of bad weather for the departure made no impression on Thomson at all. On the cold and rainy night of October 4, 1930, the *R. 101* unhooked from the mast at the airship works just as Thomson had wanted, bound southeast for its first stop, Ismailia, Egypt.

By January 28, 1986, after four days of weather holds and mechanical problems, the *Challenger's* crew was raring to go. The weather was clear, calm, and frigid after a cold front's passage. The seven-person crew climbed in just before 8:00 A.M. Among them was the winner of NASA's "Teacher in Space" competition, a public relations extravaganza that put thirty-seven-year-old Christa McAuliffe, a high school history teacher from Concord, New Hampshire, in the national spotlight. She was not the first amateur on a shuttle—Senator Jake Garn and Representative Bill Nelson had wangled joyrides before—but certainly she was the most prominent. Her students had come to the Cape to see the launch. She would be spending the mission keeping away from switches and giving live television talks on microgravity and life in space.

The other six crew members, as diverse as America itself, had their own unique stories to tell. Pilot and copilot, Dick Scobee and Mike Smith, were both trained as test pilots. Flight engineer Judy Resnik had been the second American woman in space. Air Force lieutenant colonel Ellison Onizuka was the first Japanese American to head for space, and physicist Ron McNair was the second African American astronaut. Greg Jarvis, a payload specialist from Hughes Aircraft who was going along to study fluid flow in space, would have flown on the previous mission had he not been bumped by Representative Bill Nelson.

Packed into the *Challenger's* payload bay were a tracking satellite and a one-ton comet observatory called Spartan-Halley. What was most important for NASA was not this payload, but the one to follow: NASA needed to get this flight finished so it could prepare the *Challenger* to carry a space probe called the *Ulysses*. The *Ulysses* would be studying the north and south poles of the sun after getting a gravitational boost from a close flyby of Jupiter. The *Challenger* had to go up again on May 15 for the *Ulysses* to make the rendezvous, and NASA would need every available moment after the Teacher in Space flight to get things ready.

When the "ice team" made its last checks during a two-hour hold in the countdown, the temperature of the left booster was 33°F, and the right booster was fourteen degrees colder. But with all contractors signing their approval for launch, this otherwise alarming development became nothing more than a notation in the preflight logbook. The countdown resumed and ignition took place at 11:38 A.M. It began with the lighting of each of the three main liquid-fueled engines on the rear of the *Challenger*, then both solid rocket boosters. Lit by a barrel-size igniter at the top that threw a plume of fire down the central core, each solid rocket motor came to full power in one-quarter second, sending in the blink of an eye a plume of white smoke over a fence twelve hundred feet away. The boosters' twin columns of white fire visually overwhelmed the pale blue exhaust of the liquid-fueled engines, or "hood ornaments," as booster engineers called them.

The gas from the boosters entered the exhaust nozzles at 5,700°F and left at four thousand miles per hour. A booster was the ultimate cutting torch, so hot that a steel plate near the support posts needed replacement every three or four flights because so much metal boiled away. Unseen at the time, but noticed later on film, black smoke was puffing out of the right-hand booster, at the lowermost field joint. It's likely that the hole closed soon afterward, when aluminum slag from the burning fuel stopped it up, but then opened again as the shuttle's speed rose and aero-dynamic stresses increased.

At fifty-eight seconds after booster ignition, ground-based tele-scopic cameras caught a glow from the right booster. This glow was from a jet of flame playing onto the tank and the area of the lower P-12 "sway strut" connecting the right-hand booster with the tank. The strut failed under the heat at seventy-two seconds and the booster piv-oted around the upper struts, thus shattering the external tank like a giant egg. A fireball erupted as the *Challenger* orbiter broke free and tipped sideways at supersonic speed. Never designed for this kind of shock, the craft broke up. Some fragments flew as high as 122,000 feet before falling into the ocean. One of the bigger fragments was the crew compartment, where several of the crew members remained alive—until the cabin hit the water at more than two hundred miles per hour.

THE MISSING LINK

In that long and agonizing teleconference the night before, the erratic behavior of the O-ring seals consumed so much attention that the sub-ject of exactly how a leaking booster might cause a disaster never came up. Boisjoly had been most concerned about a leak occurring in the first few seconds. Such a leak would have cut into the fuel tank, dump-ing fuel and blowing up not only the shuttle but the pad as well. But that's not what happened.

When interviewing an executive at Thiokol ten years after the disas-ter, I said that it was certainly bad luck that the leak of hot gas happened to point right at a strut connecting the external tank to the booster, clearly the worst possible place. By bad luck, I meant that if the gas leak had broken out at one of many other locations on the booster nothing disastrous might have happened. The booster with the burned-through hole in it never did burst like a punctured balloon, after all, nor had the hole grown to be extremely large even after seventy-three seconds of flight. The booster had less than a minute left to burn, and its power was already dropping off, so it was close to squeaking by when the P-12 strut failed and made the whole structure go to pieces.

"It wasn't bad luck," the executive said. According to him, the strut was the link between the engineers' generalized worries about a leak through the field joints and the catastrophe that actually happened. The P-12 strut was a stress-concentration area. This is where the gas leak broke out, and it was the first time a gas leak had ever cut through the steel case. Cold acting on the O-ring rubber, a poor joint design, and stress concentration at the struts all combined to make enough of a hole to bring the *Challenger* down.

A few minutes past 2:00 A.M. the next morning, engine operator Joe Binks climbed down the ladder from the body of the *R.101* into his power car. It was time for him to relieve the other engineer on duty, Arthur Bell. Binks looked out the window to see a steep roof in the gloom and called out in alarm to Bell, saying he had seen a church, just a few yards off.

It was the roof of the rebuilt Beauvais Cathedral, a soaring Gothic structure that had been the site of one of the first recorded failures of high technology. It was here in 1284 that either two or three 158-foot-high choir vaults collapsed, for reasons unknown, twelve years after the construction of the cathedral had been completed.

Bell couldn't hear what Binks was saying over the diesel's roar, though, and by the time he understood, the *R.101* had glided past the obstacle without a collision. The airship pitched and heaved as it left Beauvais, heading for a low range of hills called the Bois des Coutumes. The airship calmed as it reached the shelter of the rising ground. The engine telegraph called for a "slow" setting on the engines, and Bell backed off on the throttle. It was in time to hear a cry from the chief coxswain, running along a catwalk to warn everyone he could: "We're down, lads!"

The nose of the *R.101* touched the ground at low speed, skidded, then bumped up again with the recoil. Then it fell back down again, the rest of the craft settling in. At least one of its gasbags had broken open, and a hydrogen fire started in the nose and swept back toward the tail. The front of the *R. 101* lay draped across a stand of oak and hazelnut trees, with the rest of its length reaching back into a beet field. Among the wreckage were dozens of pith helmets, which had been taken along for use in the tropical climate.

Of the fifty-four men aboard, forty-eight died immediately or soon afterwards, including Sir Thomson. Among the dead was the boy-ish and excitable Sir Sefton Brancker, the director of Civil Aviation and one of the first innovators in British aviation. It was Brancker who had tried to persuade Thomson that the *R. 101* wasn't ready.

One of the survivors, Harry Leech, had been in the smoking room at the instant of the crash. The room had only two exits and both were blocked, so Leech smashed his way out through the fireproof wall. He jumped and landed in a tree. Binks and Bell waited in their power car, gaining a reprieve when a water ballast container broke above them and doused the structure. Then they jumped free, landing on the wet, soft ground of a hillside.

Disaster arose from a combination of elements, decided a court of inquiry: insufficient testing of new designs, setting out in bad weather without preparation, and loss of buoyancy at a critical moment. Said the 1931 report, "When it became important to avoid further post-ponement and the flight to India thus became urgent, there was a tendency to rely on limited experiment rather than tests under all con-ditions. . . . It is impossible to avoid the conclusion that *R. 101* would not have started for India on the evening of October 4 if it had not been that reasons of public policy were considered as making it highly desirable for her to do so if she could. . . ."

Mourning was long and elaborate. A crowd estimated at one hun-dred thousand watched the coffins leave Beauvais by train. The funeral procession arrived at Boulogne, where a British destroyer accepted the transfer. In England a special train carried them into Victoria Station. The dead lay in state at Westminster Hall, leading up to a memorial service in St. Paul's Cathedral. Prime Minister Ramsay MacDonald said that the crew's "sacrifice has been added to that glorious list of Englishmen who, on uncharted seas and unexplored lands, have gone into the unknown as pioneers and pathfinders, and have met death. . . ." During the procession from Westminster Hall to the railroad station, a half million people lined the streets. The ceremonies ended at a mass grave in Cardington.

Britain reacted by shutting down the entire dirigible program, even taking possession of the successful, privately built *R. 100* and dicing it into scrap. NASA held a memorial service for the *Challenger's* crew at Kennedy. It was attended by President Reagan and ten thousand other people. The agency responded to the *Challenger* disaster by stopping all flights for almost three years and negotiating a settlement with Morton Thiokol that had the company surrendering $10 million of profits. Thiokol worked up what has since proved to be a very successful seal for the field joints. NASA buried every last fragment of the *Challenger* under tons of concrete, using two abandoned Minuteman 3 silos, just as the agency had buried the burned hulk of *Apollo 1's* capsule.

A QUESTION OF RISK

After the *Challenger* disaster there was some anger among old-line astronauts at NASA for including members of the public on the flights. Until further notice, they reasoned, space travel was only for professionals, who were willing and able to take the risk.

I can't buy that argument. In the case of the *Challenger*, the explosion broke up the ship before anyone could lift a finger, anyway, and too quickly to send a radio message. Even if the ship had seven right-stuff test pilots on board, strapped in every one of the available seats, it wouldn't have made a bit of difference that morning. And the idea that civilians need to be shielded from danger is preposterous; more than a hundred Americans die every day in car crashes. We tolerate this astonishing carnage year in and year out as the price of mobility. Even ordinary people like me, like you, like Christa McAuliffe, should have the privilege of taking personal risks for the greater good.

If Boisjoly, prisoner of bureaucracy that he was, had been allowed the proverbial single phone call, I wish he could have summoned Dick

Scobee, the *Challenger's* commander, to the telecon. Scobee would have heard that his flight the next day was going to be an off-the-chart experiment. The *Challenger's* solid rocket boosters would be operating well outside of their tested boundaries, going farther up a sketchy but suggestive trend line of gas leaks that were becoming so common that the engineers had been finding them on almost every other flight. The crew could have heard that the flight was shaded with an element of danger simply because every flight under cold conditions showed damage to the O-rings. It would have given Scobee a choice, and the chance to insist on more information from NASA and Morton Thiokol. Crew members could have decided for themselves whether to take the chance. Putting a tracking satellite in orbit, even preparing for the important launch of the *Ulysses*, was less than a top national priority.

Both airship and spacecraft were prisoners of the many promises made to get them built: promises of safety, low cost, on-time performance, and solid technology. In hindsight, the same drive and optimism that got the work under way would look tragically misguided. According to Gerald C. Meyers, former chairman of American Motors, business managers in general avoid making contingency plans for failure. That's the kind of tack taken by losers and negative thinkers; a manager sees his job as planning for product success and continual market growth instead.

So that means the rest of us have to think about such things on our own, particularly when approached to support cutting-edge projects. One of the reasons the Apollo program had political support as long as it did was that the original budget estimates were not set artificially low (in the traditional "lowball" ploy to get the work started, then increase the budget later). The shuttle, on the other hand, started with impossible promises of cheapness and efficiency and soon got jawboned between budget cutbacks and foreseeable technical problems. The orbiter's main engines were not "just like" the Saturn's engines; the solid rocket boosters had to perform in a different dynamic environment, and the heat-protection tiles for the orbiters were radically new. Once promised to fly as many as sixty times per year, the highest rate ever achieved by the space shuttle has been eight flights a year. The "space transportation system" achieved only three flights in 1999.

This would make it easy to say that NASA has the wrong stuff, but that's just what the press was saying about the string of delays just before the final launch of the *Challenger*, ridiculing the space agency for malfunctions. We're supposed to know better by now: space travel is inherently dangerous, and so when the manager of such a program says things are not working quite right and the schedule needs to include a safety break, that should be fine with the rest of us.

There are cases in our technological history of people who stopped a project cold to get things right. After workmen had finished putting a reactor into America's first nuclear submarine, the *Nautilus*, a small steam pipe burst during dockside trials. The head of the nuclear-navy project, Hyman G. Rickover, heard that the pipe had been fabricated from the wrong material and that as a result it had no more strength than a tube you'd see as part of a guardrail. Finding further that the shipbuilder's quality-control records couldn't establish whether any more of the wrong piping had found its way into the steam system, Rickover ordered that all steam lines of that diameter—that meant thousands of linear feet—be cut out and replaced with the right stuff. According to his assistant Ted Rockwell, Rickover made a general announcement that he wanted this remembered as red-letter day, a blow for quality control that would never be forgotten. Of course it was expensive, but it sent a very clear message throughout the navy and its contractors that this guy really did value safety over deadlines. Was a costly renovation like this embarrassing to the navy? To Rickover, that would have been a stupid question. What he called the "discipline of technology" demanded nothing less.

Making the time and budget for realistic qualification tests could have saved the *R.101* and the *Challenger*. The next chapter explores why a tough round of testing—so often shortchanged—helps close the door to disaster.

4: DOUBTLESS

TESTING IS SUCH A BOTHER

Early on the morning of July 24, 1943, the U.S. submarine *Tinosa* lined up for a shot at one of Japan's largest tankers, the *Tonan Maru No. 2*. The tanker was carrying petroleum to the island of Truk. The skipper of the *Tinosa*, Lieutenant Commander L. R. Daspit, aimed four torpedoes. Two hit the *Tonan Maru* but didn't explode. The *Tinosa* fired two more; one exploded, bringing the ship to a halt. Because the tanker had no escorts yet and lay dead in the water, sinking it should have been a sure thing. Daspit ordered seven more torpedoes away over the next hour; all banged against the tanker's sides, but none exploded. As an enemy destroyer approached, Daspit fired two last torpedoes. Both hit, but neither of those exploded either. All told, out of the eleven torpedoes from the *Tinosa* that struck their target, only one had detonated.

Chalk up another big day for the notorious Mark 14, the American torpedo with the top-secret proximity fuse. Before the war began, the submarine service saw it as a superweapon, needing such security that the torpedoes shipped out to bases separately from the special fuse that was intended to increase the warhead's killing power. During the first year and a half of the war, three out of four of these torpedoes failed. By early 1943 submarine commanders across the Pacific saw the Mark 14 as one of the worst weapons of the war, so bad that they were willing to risk court-martial by disengaging the proximity fuse. That still didn't fix all the problems. The Mark 14's history "reflects discredit upon both the Bureau of Ordnance and the Naval Torpedo Station, Newport," said the bureau's chief, W. H. P. Blandy.

It's pretty obvious that your average machine should run a gauntlet of tests between prototype and rollout: Do the parts rub? Are there problems with the controls? Will it catch fire, or will a vibration shake it apart? And hundreds of other possible malfunctions. That's why we need to know how it happened that some machines went out into the world with profound flaws, so glaringly obvious later. The main reason the Mark 14 went so seriously wrong was that the torpedo station, short on budget and long on optimism, barely tested it. The few tests they ran were not of production weapons under simulated wartime conditions. In all the years of development before the war the Newport Torpedo Station fired a grand total of two armed torpedoes at a vessel, which was anchored at the time.

The second shortcoming that we'll see is failure to follow up on problem indications during tests for something else. That is, the engineers are evaluating condition A, and during this process, the readings also show an anomaly with condition B. It's like catching a glance of somebody running by as you're looking another way. Such peripheral warnings came repeatedly during the polishing of the Hubble Space Telescope's main mirror, and they help explain why no one caught the monumental blunder that sent the telescope into orbit with a misshapen reflector.

According to the press packet issued by NASA just before the shuttle launch that carried HST aloft, "Engineers used the interim period to subject the telescope to intensive testing and evaluation, assuring the greatest possible reliability. An exhaustive series of end-to-end tests involving the Science Institute, Goddard, the Tracking and Data Relay system and the spacecraft were performed during this time, resulting in overall improvement in system reliability."

But it wasn't true, not to the extent that the NASA release suggested the mirror was well tested. The mirror sat in storage with a grievous error during the eight years between its completion and eventual launch. The space telescope went up with a rudimentary kind of error that amateur astronomers check for as they grind and polish a mirror on a workbench in the basement.

With the Hubble Space Telescope, NASA sustained a public-relations hit that was exceeded only by the *Challenger* disaster. For a long time to come the name Perkin-Elmer will be remembered by astronomers as the one that caused Hubble all the trouble. After two reorganizations the company now goes under a different name. These failures to test were in the end extremely costly in dollars and damaged reputations. Before the Mark 14 torpedo failures, the Newport Torpedo Station was the country's only torpedo development center, which is just the way Rhode Island politicians wanted it; but after the war the station was shut down and then razed for hotels and condominiums.

THE RACE

Think of a good testing program simply as a race between inventors and users: who's going to find the bugs first? It was clearly the users, in the case of the Mark 14 torpedo.

In 1869 the navy bought the land for the Newport Torpedo Station: Goat Island in the harbor of Newport, Rhode Island. Newport was jealous of its status. The local congressional delegation fought off all attempts to move jobs out of the plant in the interest of competition or efficiency. But though Newport kept its monopoly on naval torpedo research, it was at best a stalemate, because Congress was willing to put up little cash for new weapons in the doldrums between the great wars.

But in Newport's heyday one of the new weapons was to be the Mark 14 torpedo for submarines. It was the brainchild of Ralph Waldo Christie, a product of the submarine school at New London and a skilled mechanical engineer. Powered by a steam-driven turbine, propelled by compressed air and alcohol, the torpedo carried a quarter-ton of explosives. The breakthrough was a proximity fuse designed to set off the explosive charge as the torpedo passed under a warship. The fuse would know it was under a ship by a change in the earth's magnetic field, caused by all the iron in the ship's hull. Said Christie, this would be much more devastating than a conventional torpedo that ran shallow, rammed a ship's side, and blew up only on contact, because ship designers were busy adding armored belts to ships for attacks from the side. They were not protecting warship hulls from blasts underneath.

In 1922, Christie started working up a proximity-fuse prototype. Two years later he had a functional one that could fit on a torpedo. He began hounding the navy's Bureau of Ordnance for steel ships to shoot and sink. Two more years later the navy let Newport have one old submarine bound for the scrapyard. After one of the two torpedoes went off and sunk the anchored target, that was it: the navy liked the results so much that they offered no more ships for him to test it on, not unless Newport was willing to pay the salvage cost if they sunk one. So there were no more live tests of the Mark 14 until after World War II started, fifteen years later. By that time thousands of Mark 14s had been manufactured.

Newport had fired experimental Mark 14s over the years, but convenience, caution, and rigorous economy ruled. The torpedoes didn't have explosives on board because that might blow them up. They didn't ram solid objects to test the contact detonator because that would damage the delicate machinery. When used out at sea in war games, sailors knew to keep the depth setting so far below the surface that the torpedoes wouldn't actually strike a ship and damage themselves. During tests at the station, the same torpedoes went out across Newport Harbor, again and again, floating to the surface after the run for recovery and reuse. The result: with time and tinkering the test torpedoes grew significantly different from the ones going into inventory.

By 1937, Newport had three thousand men at work, apparently all of whom prided themselves on slow and careful craftsmanship. But even at the rate of only two or three torpedoes a day out of the whole factory, the fish began piling up, each one too precious to pull out of inventory and test under warlike conditions.

American submarines began firing Mark 14 "war shots" in early 1942. First reports showed them passing under targets without going off. Commanders said they must be running too deep. The Bureau of Ordnance refused to believe the reports until the summer, eight hundred war shots later, when Vice Admiral Charles Lockwood improvised his own test in Frenchman's Bay, on the Australian coast, with a fishing net. Firing live torpedoes through the mesh and measuring the holes they made, he proved that the Mark 14 was consistently running more than ten feet deeper than the setting dialed in by the submarine crew. The reasons were many: a problem with the depth-keeping mechanism, an instrument out of calibration back at Newport, and the failure to take out random real torpedoes from the stockpile and test them for depth control.

After the Lockwood tests, the Bureau of Ordnance sent a message out to adjust the torpedoes to run at a shallower depth. Submarine commanders soon reported back with much rancor that the depth problem was solved but that most of the torpedoes they fired were either going off too soon or, more typically, not exploding at all. Totally exasperated by the first problem and figuring that no one at headquarters would pay attention to the second, some submarine crews began deactivating the proximity fuse even though this was a serious offense. In that way, commanders would depend on the backup detonator instead, the old reliable, the contact fuse that would trigger the warhead when the torpedo's nose rammed an enemy ship.

Now it was time for the torpedoes to reveal their third and final untested flaw. The contact detonator on the nose, a fairly idiot-proof device, didn't work, either. Imagine the reaction of crews who had spent weeks of miserable submarine life to get into a good firing position, letting off a brace of torpedoes, then hearing them thud against the side of a ship without exploding.

Lockwood got to work again in August 1943, firing live torpedoes at Hawaiian cliffs and dropping disarmed ones onto angled steel plates from a crane to see what the problem was. It was the design of the detonator: every torpedo that struck a target head-on, or nearly so, was not going off because the impact had distorted the mechanism. In the end it took more than two years of bitter experience before American submarines had in their possession modified Mark 14s that were good enough to use in battle. Had Newport been allowed to, or really wanted to, test its production torpedoes it could have found and fixed the problems years before. Newport failed to test their fish because the budget shortage fit well with their predisposition to see their finely crafted machines as too valuable, and too good, to waste on tests.

FATAL CONFIDENCE

Certainty is a good thing, up to a point. While looking through a marketing magazine once, I saw an advertisement with the maxim Never Doubt Your Beliefs, and Never Believe Your Doubts. It's excellent advice for salesmen who meet tough resistance from customers every day, but I don't want a person who adheres to this maxim to be in charge of deciding when to put a new machine into service.

The contemporary insult "clueless" is directed at people we see as completely out of touch with important events. If we adapt that word a little, we come up with the new insult of "doubtless," meaning a person who is rock-solid certain he knows all he needs to know and believes that nothing can go wrong. Being doubtless is as dangerous in its own way as being clueless.

One of the people who knew the importance of testing equipment without mercy was Hyman G. Rickover, who immigrated from Poland at age four. He graduated from the U.S. Naval Academy in 1918 but held only one ship command in his life. During World War II, Rickover got the assignment to make sure that warships' electrical equipment could survive battle conditions. This wartime work set the course for the rest of his life: working harder than anyone around him, abusing and belittling, making manufacturers cringe, yelling or even screaming into telephones, taking nothing for granted except his own integrity. As chief of the electrical section at the Navy Bureau of Ships he traveled at least one hundred thousand miles a year, trying to inspect every battle-damaged ship that arrived for repairs to see what held up and what broke up.

It was a tireless effort on behalf of the men whose lives could well depend on electrical gear that would not fail. But by September 1947 the career of Captain Hyman Rickover appeared kaput, along with those of thousands of other postwar officers. When he asked his bosses for permission to begin developing nuclear power for submarine propulsion, it was hard to take this junior officer, this technocrat before his time, seriously.

Experts at the new Atomic Energy Commission told Rickover that submarine propulsion was so difficult—because of the constricted space and the high temperatures and great pressures needed—that the subsurface nuclear navy would probably be the last venue to get a power-producing reactor. Reactors of the era were little more than stacks of graphite bricks and fuel pellets, good only for making plutonium from uranium. The navy scattered Rickover's small group of assistants across the country. Given the powerless title of "special assistant for nuclear matters" and left without staff or budget, Rickover moved into the remodeled women's restroom that the Bureau of Ships assigned him for an office.

But Rickover prevailed, partly by the brilliant move of creating and heading up a "naval reactors branch" at the AEC and a "nuclear power branch" at the navy. If the navy balked at something he would challenge them as a representative of the AEC, and vice versa. Submarines turned out to be the first practical use of nuclear power after all, and sooner than anyone else had predicted. By 1960, this bureaucrat and his cadre of engineers had stormed every pocket of resistance and bullied manufacturers into putting together a successful submarine using exotic materials like hafnium and zirconium. The *Nautilus* was capable of submerging at Pearl Harbor, transiting the North Pole, and popping up on the East Coast. The voyage was as much of a shock to the Soviets as their *Sputnik* satellite launch was to us.

Admiral Rickover didn't simply invite bad news; he demanded it, every weekday and weekends, too, much to the fury of naval contractors who claimed that Rickover's roving shipyard inspectors exaggerated problems just to keep their boss happy. Rickover personally interviewed thousands of officers for the subsurface nuclear navy, snapping questions and challenges at them to test their resilience. Unless he was seriously ill he embarked on every nuclear submarine's initial sea trials, figuring that if a badly brazed piping joint broke at test depth he should be on hand to accept the consequences. Handed a new piece of electrical gear proposed for use on a submarine, he was likely either to throw it out the window or to hammer on his radiator with it, testing for toughness. The naval reactors branch he built and directed in the navy stands as the most quality-driven program in the world. The mention of his name in the submarine service still sets off hours of Rickover stories.

TEST TO DESTRUCTION

Certainly some people of the nineteenth century knew the importance of rigorous trials. Consider how inventors of the typewriter learned to sacrifice on the altar of uncompromising tests. There were three men present at the conception of the first successful typewriter: Christopher Sholes, who came up with a way to space the letters; S. W. Soule, who conceived the breakthrough idea of swinging typebars that would pivot inward to land on a single target area; and Carlos Glidden, who rounded up the money. By the following September they had put together one machine featuring a piano-style keyboard. They used that contraption to create what may count as the first junk mail in history, sending form letters to friends and relatives in an attempt to raise money. One of the recipients was James Densmore, a promoter. Densmore agreed to pay all expenses to date, in return for a quarter interest. Then two of the men dropped out, leaving only Densmore and Sholes in the quest.

Sholes kept himself busy building intricate prototypes out of metal pieces that he crafted with painstaking care at a small forge. With each model he labored to fix problems detected on previous models. The partners began shipping the precious machines out to working stenographers for them to try. After a while, Sholes was greatly perturbed to hear that one of the stenographers, James Clephane of Washington, D.C., had tested and broken every machine they sent him, soon after its arrival. Sholes was going to stop this outrage, but Densmore convinced him otherwise. Clephane was a stenographer for a federal court and needed a typewriter to turn his shorthand into clean copy. He was the kind of unforgiving expert user the typewriter would be meeting, and if Sholes's best handiwork could not withstand the full fury of Clephane's typing, they might as well close up the business. Mr. Clephane, said Densmore, was a blessing to the enterprise and not a curse.

At the core, the uppermost reason for not testing is a mindset that everything leaders have ordered has been done exactly as prescribed. This is a remarkably naive presumption, as the history of technology, and construction in particular, is packed with examples of workers who upon finding a tool or design to be inefficient have, in effect, said, I spit on your stupid ideas, and switched parts around or kept modifying them until they fit. Sometimes they told the designer or engineer about it; sometimes they did not.

TROUBLE AT THE EDGE OF THE UNIVERSE

The fact that conception and construction don't always match up played into the early, maddening years of the Hubble Space Telescope. The key mistake wouldn't have made much of a difference had the manufacturer and NASA been alert to what certain tests were telling them. But because those tests were revealing a subtle problem outside of the exact scope of the tests, and because the problem was too unbelievable for words, nobody followed up.

On May 20, 1990, the $1.5 billion Hubble Space Telescope took its first picture, of a star cluster in the constellation Carina. Over the following five weeks the press waited for photographs with unconcealed impatience. NASA had promised a rich harvest of astronomical discoveries and spectacular photographs through the Wide-Field/Planetary Camera. "NASA's Incredible Time Machine," the public affairs office called it, claiming that it would be able to look back to the earliest days of the big bang. It would extend our view of space from two billion light-years through ground-based telescopes to at least fourteen billion light-years. Whether or not most of the reporters knew that this was hogwash—ground-based telescopes were already picking up faint objects at the edge of the visible universe, and besides, no telescope can see out more light-years than is equal to the fifteen-billion-year age of the universe—all of them wanted news of what Hubble could do.

Then on June 27, at a press conference at Goddard Space Flight Center's visitor center, the press assembled and heard something else entirely. Two days before, on a day remembered as Black Monday, NASA had finally agreed with astronomers that the space telescope suffered from a serious flaw in its main mirror. First detected by English mathematician Christopher Burrows just ten days after the telescope opened its lid for first light, HST had a bad case of "spherical aberration" that threw off the otherwise precise focus to be expected from a large, superbly polished reflecting telescope.

Reporters had been hearing rumors of some kind of problem with HST, and at the press conference they heard the worst possible interpretations of the news; it was like spin-doctoring in reverse. "No real science can be done with the telescope's main camera," said NASA's program scientist for the project, referring to the Wide-Field/ Planetary Camera, the WF/PC that everybody figured would be bringing in the color photos that so pleased the taxpayers.

The news came across in such a way that a casual listener might jump to the thought that NASA might as well have put a railroad tank car in orbit, which was about the same size. The crisis wasn't quite that bad. In time scientists worked out computer corrections that would let the telescope gather good images but at the cost of discarding most of the mirror's light-gathering power. But the telescope would be gravely limited until the first repair mission more than three years later.

In the same press conference, NASA officials took direct aim at the maker of the Hubble's main mirror. It was Perkin-Elmer Corporation of Danbury, Connecticut, by now bought up by General Motors and renamed Hughes Danbury Optical Systems (since renamed again, to

Raytheon Danbury). Once Perkin-Elmer had been world famous for its precision optics. It had been a favored contractor for the air force's spy-satellite program. NASA said federal investigators would be moving into all plants that had ever worked on the main mirror. They would gather up all relevant documents, and they would seal off or impound all the measuring and polishing equipment. Soon instruments and workers spilled the beans, making Hubble one of the easier technological breakdowns in history to figure out.

The project officially began in 1965 when NASA started planning the "large space telescope." Gliding above nearly all Earth's atmosphere, a space telescope would have resolving power better than any ground-based telescope feasible at the time, whatever its size. But such a telescope would need maintenance in space, not then practical, and so it remained a concept until the approach of the space shuttle. Congress approved real money for it in 1977, with NASA promising to have it ready for a December 1983 shuttle flight.

NASA's Marshall Space Flight Center would assemble the superbly sensitive components. The main mirror would be the heart of HST. Contractor Perkin-Elmer would grind and polish the mirror, accurate to millionths of an inch. Nearly eight feet across, the silvered mirror would deliver ultraviolet and visible light to a battery of five scientific instruments via a secondary mirror.

GRIND IT DOWN

In December 1978, Corning Glass delivered the flat slab of "zero expansion" glass to Perkin-Elmer's plant in Wilton, Connecticut. The equipment at Wilton ground off two hundred pounds of glass, hollowing out a rough concavity that turned the slab into a very shallow bowl. The mirror went by truck to the main plant in Danbury for fine polishing in May 1980.

Perkin-Elmer was famous for its workers' skill at precision optics and had made big mirrors before, but the company believed this job would require something special. The Danbury plant set up a computer-

operated polishing wheel that whirred across the face of the mirror for
hours on end. Every few days technicians halted the polishing equipment
and pushed the mirror on a set of rails into another room for inspection
of the hyperbolic curve. Perkin-Elmer had used the technique for making spy-satellite mirrors, but this was the first time its special instrument—
the reflective null corrector—would be used for a scientific satellite.

The reflective null corrector was a precision-built instrument for
measuring the curves on large mirrors. Perkin-Elmer used other types of
null correctors for aspects of the project as well, but the supposedly super-
accurate one that caused the problem was about the size of a barrel. This
reflective null corrector had two mirrors and a lens set into a frame. Every
time they checked the mirror's curvature, technicians beamed a light
through the center of the instrument, down to the polished surface of the
mirror's glass. The light beam reflected off the face of the mirror and back
into the null corrector. That made a pattern of black-and-white "interference" lines that optics experts at the plant photographed for study. The
shape of the lines around the face of the mirror told them how to adjust
the computer's polishing path for the next run.

So what went wrong? The setting of the reflective null corrector
(see Figure 5). Perkin-Elmer had originally used the corrector for a
smaller test mirror. Part of the changeover process for the Hubble's
larger mirror required technicians to insert an ultraprecise metal measuring rod (called the "B Rod") to make sure they put exactly the right
distance between the lens and the lower mirror in the instrument. If
things had worked as intended, a light beam would have passed
through a hole in a cap on the end of the B Rod at a precise spot,
reflected off the rod's end, and reached a detector, signifying that the
lens and mirror were precisely the right distance apart.

But the light beam did not go through the tiny hole, as it was
supposed to. Though the cap in question was painted flat black to prevent stray reflections and make sure that the only light reaching the
detector had to pass through the intended hole first, a bit of paint had
flecked off the metal on the cap. The light reflected off this bright spot
and sorely baffled the technicians: the lower mirror was supposed to be
in the right place, but when they tried to move the lens to where it

FIGURE 5: REFLECTIVE NULL CORRECTOR
(FOR HUBBLE SPACE TELESCOPE)

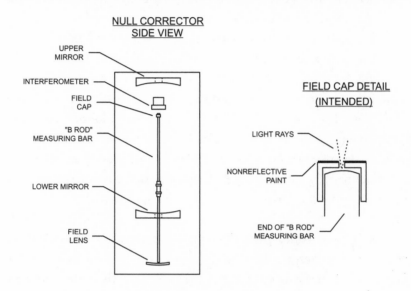

NULL CORRECTOR
SIDE VIEW

UPPER MIRROR
INTERFEROMETER
FIELD CAP
"B ROD" MEASURING BAR
LOWER MIRROR
FIELD LENS

FIELD CAP DETAIL
(INTENDED)

LIGHT RAYS
NONREFLECTIVE PAINT
END OF "B ROD" MEASURING BAR

PROBABLE SEQUENCE

1. Reflective null corrector (RNC) is relied on to be only guide for grinding curve of Hubble's main mirror.

2. Adjustment of RNC for job depends on use of precise metal measuring rod, "B Rod," to set distance between field lens and lower mirror's center of curvature.

3. Final measurement of B Rod involves light ray bouncing off end of rod through field cap on end of rod.

4. Instead, light bounces off end of field cap, where nonreflective paint has flaked.

5. Field cap error makes B Rod seem 1.3 mm longer than it is.

6. Technicians add metal washers to finish adjustment.

7. Error causes spherical curve in part of Hubble's mirror.

Adapted from NASA

should go it wouldn't move far enough. The technicians were pressed for time, because final polishing of the Hubble mirror could not start until the null corrector was ready. The technicians had been told in very clear terms to get the job done, and they had been assured that everything about the setup was infallible.

The three technicians talked the problem over among themselves. Inspiration struck: if they made the null corrector's lens move just a little bit more, everything should fit. So they inserted three flattened hardware-store washers into the reflective null corrector, bringing the lens 1.3 millimeters lower than intended. Nobody else knew of the change, nobody tested the null corrector before it went into place in the measuring tower at Danbury, and nobody discussed the field modification with Lucian A. Montagnino, the temperamental but brilliant engineer in charge of the reflective null corrector conversion project.

As we'll see with *Apollo 13* in chapter 8, such off-the-book shortcuts have been responsible for a long history of disastrous consequences, with the damage rising in correlation to the complexity and power of the machine in question. Even before reaching full power during its initial tests, the Fermi breeder reactor at Lagoona Beach, Michigan, suffered a partial flow blockage of its liquid sodium coolant because a zirconium sheet tore loose at the base of the reactor. Fermi's builders had added the sheet as an afterthought to protect another part, not considering how it might fail or even making note of its existence on the design plans.

MAKE IT FIT

A Viscount airliner lost its wing and crashed in England in February 1958 because a mechanic had decided to saw off a metal pin so as to make a new part fit. At the time he was repairing the "elevator" control surfaces that make the craft climb and descend. To the mechanic, the long pin for the elevator spring tab was obviously a manufacturer's error and so needed fixing, but to the manufacturer the pin was a foolproof method to keep mechanics from confusing one similar-looking part with another. The consequence of the mechanic's initiative, forcing the wrong part to fit, was to reverse the mechanism's workings. It turned the airliner into a machine that fought the pilots' every attempt to control its pitch. After the inspector approved the job, a crash was inevitable.

Incredibly, at Perkin-Elmer this aspect of null corrector assembly didn't rate an expert inspection, even though it was a core function that affected the entire telescope. Taking great care not to jar anything, the men raised the maladjusted instrument into the measuring tower so it could begin work watching over the polishing of Hubble's main mirror. Over the next year of polishing, the bad information from the corrector gave the Hubble mirror a dual personality. Although the innermost, central area on the primary mirror had the proper hyperbolic shape at the end of the process, the rest had been ground down a bit too far, giving it the curve of a sphere. This "spherical aberration" meant that light rays didn't come to a sharp focus. This flaw gave photographs of stars a strange look, with every star having a bright central area and then halos of light surrounding it, as if every star in the universe had suddenly gone supernova.

Computers could restore sharpness by pulling out just the central bright core of each star, discarding 85 percent of the light in the process, so the final effect of spherical aberration was to make the Hubble mirror equivalent to one much smaller. The repair mission in 1993 vastly improved the flawed mirror's usefulness but could not fix the entire problem.

It's very unlikely that the consortium of Eastman Kodak and Itek Corporation, which had competed with Perkin-Elmer to win the NASA contract to make the Hubble main mirror, would have delivered a defective mirror. In its bid Kodak had proposed to grind two mirrors, cross-testing them to pick the best one for the telescope. But Kodak had wanted $105 million to build the main mirror, $35 million more than Perkin-Elmer's bid, and so Kodak lost the competition. In time, Perkin-Elmer's costs climbed so high (its costs eventually totaled $440 million) that NASA refused to pay any more.

Perkin-Elmer knew what to do about that. As soon as possible after applying the main mirror's reflective coating, the company began shutting down the work on a crash schedule, which included canceling the thorough optical test originally planned for December 1981. This test, which was strongly recommended by an independent review earlier in the year, would have caught the problem. But time had run out. The work closed down so rapidly, says Eric Chaisson in his book *The Hubble Wars*, that when investigators arrived at the Danbury plant to take names they found the mirror polishing laboratory preserved like a large time capsule. The technicians had walked out nine years before and abandoned everything in it. Coffee had evaporated over the years from the cups, leaving a brown crust.

According to investigative reporters Robert Capers and Eric Lipton of the *Hartford Courant*, Perkin-Elmer employees received hints well before the project's final days that something was wrong. Because the information had come from the periphery, it was disregarded by the key people. The first sign that the main null corrector was off came before the mirror's final polishing began at Danbury, when two null correctors gave different results. No one paid attention to that glitch because the mirror was still rough. Later, an alignment instrument showed a mirror problem. The final indication came on May 26, 1981, when another test instrument checking for the center of curvature revealed spherical aberration. Once again, no one was alarmed, because they believed that the main null corrector was much more precise than the conflicting instrument, and the test wasn't intended to check for this problem, anyway.

An old slang phrase for a worn-out airplane is a thousand spare parts flying in close formation. It also describes one of the causes of Hubble's mirror crisis. According to Eric Chaisson, the project kept ten thousand people very busy for most of a decade, but little effort went into coordinating the efforts of these people with the expertise of the community of astronomers the telescope was built to serve. If it had been otherwise NASA would have insisted that the bid include a complete systems test before launch. The position of Perkin-Elmer and NASA afterward was that a complete "end-to-end" optical test would have been prohibitively expensive, and that's why it wasn't done. It's true that a full end-to-end test would have been very expensive, but a much less elaborate test could have caught the main mirror's problem.

It was in many ways a repeat of the dynamic behind the *Apollo 1* fire in 1967. This from a speech that flight director Gene Kranz gave to Apollo workers at Houston afterward: "We were too gung ho about the schedule and we locked out all the problems we saw each day in our work. Every element of the program was in trouble and so were we. . . . Nothing we did had any shelf life. Not one of us stood up and said, 'Dammit, stop!' "

Decentralization is good politics, because it spreads the work out, and that keeps the maximum number of businesses and communities happy. One result: low-level NASA staffers in the field had real problems with the quality-control procedure they saw at Perkin-Elmer but felt they had no support from above to do anything about it.

THE HARD ROAD TO MARS

A similar failure to pick up on a peripheral warning culminated in another NASA embarrassment in September 1999, after the $125 million *Mars Climate Orbiter* dipped too low when making its first aerobraking pass through the Martian atmosphere. The craft either broke up or went out of control, vanishing into a stealth orbit somewhere around the sun. Signs of the problem began to appear five months before the arrival at Mars. During the trip the craft had been firing its thruster at least ten times more often than planners had expected would be necessary to correct tiny errors from its onboard flywheels. The second warning came from Doppler measurements of the spacecraft's velocity, which suggested that the flight path was a little off.

The problem actually resided back on Earth, in an erroneous computer file used to process information arriving from the spacecraft about the thruster's correctional burns. The output of the computer file (called "Small Forces") should have been in metric units called Newton seconds but was instead in pound seconds. It threw off the navigational software.

Three months later, the *Mars Polar Lander* crashed on the planet's surface. The braking engines cut out when the craft was still 130 feet up, turning it into a "Crash Lander." According to the investigation, it happened because jolts from the midair deployment of the three landing legs fooled a sensor into thinking the spacecraft had touched down and therefore the engines should stop. The root cause was a small programming error, missed during ground tests because the checks were done at a time when the sensor was wired wrong. Although technicians later repaired the sensor wiring, it didn't occur to anyone that the tests had to be redone afterward.

There can be a human cost following a probe's failure, even though nobody gets hurt. Take Michael Malin, who helped build the camera system for the earlier *Mars Observer* mission. In August 1993, three days from its destination, the *Mars Observer* dropped out of radio contact forever. Malin told a *Science* reporter afterward, "A TV reporter will go up to someone who's just lost children in a fire and say, 'How do you feel?' Well, that's how I feel." So Malin went to work building cameras and control systems—for what? The equally ill-fated *Mars Climate Orbiter* and *Mars Polar Lander* missions.

"These things just keep happening," said Marcia Smith, who specializes in space issues for the Congressional Research Service. "We don't seem to be doing any better." It capped off a truly terrible public relations year for NASA; only three shuttle missions got off the ground in 1999, a new low since the *Challenger* era.

THE SAFE TEST

Sometimes people hold back on doing extreme tests out of a seemingly benign impulse, namely the desire to protect the lives of test pilots. If a test under extreme conditions might be too dangerous, the testers get to do it the safe way instead. This attitude may have played a part in the following chain of events. When Boeing was doing the certification testing for its twin-engine 767 airliner, the Federal Aviation Administration (FAA) required Boeing to report back on what would happen if some goof-up triggered a "thrust reverser" in flight. Reversers are basically power-driven buckets that swing out and force a jet engine's exhaust to go forward, acting something like a retrorocket on a spacecraft. Pilots use thrust reversers only on the ground.

Was the FAA being compulsive about this? Not really. Older jets had a mechanical safety that allowed the thrust reverser to engage only when the throttle was idling, but newer jets like the 767 had an electronic approach to safety interlocks instead. One of the key safety devices on the high-tech system was an "auto restow" feature. A computer continuously gathered information from electronic sensors about the status of the thrust reversers. If the sensors warned the computer that the reversers were sneaking open, the computer would direct hydraulic fluid through a "directional control valve" and force the reversers shut again. At the same time a light— labeled "Rev Isln"—would come on in the cockpit to warn the pilot that the computer was forcing the thrust reverser closed. It was a little like having a device bolted to the outside of your car to monitor whether your doors were coming open and, if so, to slam them shut automatically.

So the FAA wanted to know what would happen to the airliner if a thrust reverser came on through some problem in flight and if the auto-restore safety interlock didn't work as intended; never mind Boeing's insistence that it would never happen. But the FAA did allow Boeing to set up the conditions of the 1982 test. Boeing's rather benign midair test worked this way: First the pilots slowed the airplane to 250 knots, then idled the power on the engine to be tested. After the engine had spun down they forced the thrust reverser on that side to deploy. The off-center thrust forced the airliner to swerve in that direction, but the pilots had no trouble maintaining control.

The test pilots reported the good results, and the FAA accepted this single test as conclusive on the subject. Boeing used the in-flight information to program its training simulator, so every pilot who used the simulator and was confronted with this failure knew just what would happen. If the unlikely came about, the airplane would do nothing more serious than swerve in midair.

Nobody told the airplane. On May 26, 1991, a Lauda Air crew was flying a 767 over Thailand, climbing at high power, five minutes out of the Bangkok airport on its way to Vienna. The crew noticed the "Rev Isln" light coming on, first yellow and then green. Green was the signal that the reverser was engaged. It was true, at least on the left engine. One likely scenario involved electrical malfunctions, combined with the directional control valve operating incorrectly. In trying to restow the thrust reversers, instead the safety interlock ordered the reversers on the left engine to open. Four seconds later, with much of the lift on the left wing canceled out, the airplane flipped into a violent twisting dive to the left. A half minute after the first flicker, the 767 broke up in a screaming dive.

The point here is not that a thorough test would have told the pilots Thomas J. Welch and Josef Thumer what to do. A thrust reverser coming on in flight may not have been survivable, anyway. But a thorough test would have informed the FAA and Boeing that thrust reversers coming on in midair was such a dangerous thing that Boeing needed to install a positive lock that would stop it from ever happening. And that's what Boeing did after the Lauda Air crash, 223 lives later.

There have been people, though, who run their machine to the edge of destruction, and sometimes beyond, because they want to be sure of how it will react at the redline. Something like people who discover under great stress or motivation that they are capable of great things, these experiments have revealed that machines sometimes tolerate much more abuse than anyone imagined they could. You might think as I did that no ordinary jetliner can go past the speed of sound without breaking up into tiny supersonic fragments, but in August 1961 a Douglas Aircraft test pilot took his DC-8 to fifty-two thousand feet, put it in a power dive over Edwards Air Force Base and broke the sound barrier without dying along the way.

THE SHUTTLE'S FINAL OPTION

Some machines cannot be safely tested or even understood with computer simulation, since too little of the basics are known and nobody knows how to gather the data. One such dilemma faces the space shuttle program right now. It has to do with the launch sequence for the shuttle. We know that the main engines on the shuttle are stressed to the max and can shut down in flight. One main engine shut down during climb out on a flight in July 1985. During the launch of the STS-93 mission on July 22, 2000, three engines on the *Columbia* came close to shutting down early enough to cause an emergency landing, because of a hydrogen leak caused by punctured piping in the right engine at the time of launch.

If two or more engines shut down during the latter stages of climb out the shuttle could land at abort bases in Spain, Senegal, or Morocco. But if it happened earlier the shuttle would have to try the "return to launch site" maneuver, where after dropping its boosters the shuttle would flip around, engines still firing, using the leftover fuel to reverse directions. It would actually fly backward for a time through the upper atmosphere, shrouded in very hot exhaust gases, in an attempt to get back to Kennedy's runway. Astronauts have never tried the maneuver in a real shuttle—too risky—and it can't be tested in a wind tunnel, either.

Tests can be difficult in other situations, too, but maybe we need to think of testing as something broader. Based on her years of visits to nuclear power plants as part of an international safety program, anthropologist Constance Perin argues that people who own complex or risky machines need a new attitude. They should be treating their operators as field scientists who can provide unvarnished information about how these systems actually work, or don't work, under a wide variety of conditions. With such an attitude, Perin says, managers would know to welcome bad news rather than avoid it. Their world would be a laboratory, with many surprises left to uncover.

I like to think of thorough testing as a badge of confidence: confidence that the machine can take abuse, and if it breaks, the designers will be able to find a solution. And it's good business, reassuring users that there's no place they can go that the manufacturer hasn't been already.

5: THE REALLY BAD DAY

PANIC AND TRIUMPH ON THE MACHINE FRONTIER

Strange how it is that one person's bad day can be darn close to a miracle for other people in the same place. Take Mrs. Al Kaminsky, a passenger on American Airlines Flight 96 in the leg between Detroit and Buffalo. The date was June 12, 1972. First the flight attendants refused her request to open the cocktail lounge at the rear of the DC-10 jumbo jet, saying that because Detroit was a dry town it was unlawful to open the bar on the ground there, then telling her in the air that the trip wasn't long enough to open the bar there either. Then the rear cargo door blew out with an almighty bang, dropping the unoccupied bar through the floor and causing panels in the cabin to fly open. One of them in crashing open gashed Mrs. Kaminsky's face. Then when the plane landed and she was hesitating about jumping from the top of the escape slide at the airport, the flight attendant pushed her out. Mrs. Kaminsky slipped at the bottom and gashed her foot to the bone. Finally, a gang of FBI agents detained her in Detroit so they could interrogate her and her husband and all the other passengers as suspected terrorists, for God's sake!

"They tried to kill me!" she was overheard saying at a pay phone later.

Yes, if "they" means an armload of thoughtlessly designed parts on the DC-10's cargo door. But more important is that some people, mainly Captain Bryce McCormick, saved her life. It was a very close call, indeed. When I met Flight 96's chief flight attendant, Cydya Smith, years later at Grand Central Station in New York City, I had the strangest impression that I was shaking hands with a statistical ghost, a person that should not be alive now. In every other instance in which airliners in flight faced the kind of mechanical crisis that Flight 96 did—pilots losing most of the flight controls—those airplanes all crashed, killing either everybody on board or many of them. McCormick, who had mentally girded himself two months earlier for just such an extreme emergency, brought everybody back home. He's since died (of natural causes), but after talking with his wife about the man he was, I'm convinced that he prepared for just this sort of emergency out of a sense of deep personal responsibility. Not responsibility to a machine, but to the people who depended on its safe workings.

STRETCHING THE ZONE

You may have heard someone talking about some kind of "worst-case scenario," but they probably don't mean that literally, since nobody's going to make it out alive from a true worst case. Rather, they're referring to a situation that lies within the zone of survivability. In the case of helicopters, a true "worst case" would be the main rotors stopping, a disaster that happened to a U.S. Navy Seawolf helicopter gunship over Dung Island in Vietnam after its gearbox was hit by a burst of 12.7 mm antiaircraft fire. The freak incident locked up the rotor hub, and the machine fell freely a thousand feet to the ground. Since a main rotor failure is not survivable, helicopter pilots don't bother to train on how to handle that scenario. But helicopter pilots do learn how to handle a really bad day, and in this way they extend the edges of their survivability zone. For example, pilots wanting a license as flight instructor have to learn autorotation, which is the skill of landing a helicopter with only the leftover momentum of the aircraft and its rotor blades after its engine has shut down.

Too often we just go through the motions of preparing for bad situations, assuming that the only contingencies that might happen are simple or convenient ones. One of the missed opportunities at Babcock & Wilcox's simulator training center before the TMI-2 crisis was that the company only lobbed easy softballs to their reactor operator trainees, meaning they presented them with only textbook problems, which were so simple that automated controls could handle them without human intervention. These were single-cause problems, signaled clearly by the instruments. When much more vexing multiple problems came up in the real world, operators were not ready.

We need to accept that on really bad days, more than one thing is going to go wrong. It will be maddening and frightening. There is no law of the universe that says one bad thing cannot be followed immediately by several more, and even worse, things. Some of the emergencies arising are so fraught with danger, such as the space shuttle's "return to launch site" maneuver, that training people to survive it would be about as hazardous as going through the real thing. Still there is virtue in honesty, in telling workers that someday they might have to face a rare type of crisis for which their responses can only be discussed, not practiced, beforehand.

The bad things might be strung together by cause and effect or they could emerge at the same time out of hidden problems no one had noticed. The causes of *Apollo 13*'s near disaster fell in the latter category. But those three astronauts survived, in part, because engineers had already written up a way to get a crippled Apollo module home with an "LM Lifeboat" scenario, in which the lunar module's landing engine would provide thrust.

COOL HEAD, KEEN EYE

The story of how Captain Bryce McCormick happened to be in the wrong place at the right time is one of the most remarkable stories from the machine frontier. The really bad day he had planned for was a little different than what actually happened, but it prepared him well enough.

In 1940, Bryce McCormick was married, twenty years old, and working at an aircraft factory in southern California when he decided to pursue his dream of becoming an airline pilot. He talked Bank of America into loaning him the fee for his flying-school training, and he was so earnest about it that he moved to Arizona, leaving his wife in San Diego for the duration of the course. McCormick applied to be an Army Air Corps pilot after finishing the course but was turned back after a mistaken diagnosis of a heart condition, which the army medical staff would not change.

McCormick worked as a civilian instructor flying Stearman biplanes until the defense emergency made all the private flying schools in his area shut down. Seeing his options as a pilot dwindling, McCormick walked into the American Airlines office in Burbank and asked for a job. The interviewer was about to run him off when as a last question he asked McCormick if he had anything special in his background. Just one thing, said McCormick; he had been the youngest Boy Scout ever to win the Eagle award in the state of Kansas. That's different, said the interviewer, taking a sudden interest. He sent McCormick off for a physical. American Airlines found McCormick's health to be perfect and hired him immediately. McCormick spent the rest of his career with the airline. A man of considerable energy, in the early years he would return after duty to his family in Palos Verdes Estates, California, change his clothes, and immediately resume his hobby of building a house, on a terrace overlooking the ocean. As he laid the blocks with a copilot friend, his wife, Bonnie, struck the joints.

During the first twenty-eight years with American, McCormick mastered six types of airliners, four of them built by Douglas Aircraft. And now a fifth Douglas model was coming up, the DC-10 jumbo jet, assembled by McDonnell Douglas, a company created in 1967 when an aerospace manufacturer acquired Douglas Aircraft to pull it out of financial trouble.

McCormick had the chance to examine the DC-10 early on. He looked it over tip to top, even climbing inside the cargo compartment under the passenger cabin. One thing he saw in there bothered him: the location of the control linkages that governed the rear engine, rudder, and elevator on the tail. The DC-10 had three independent sets of control cables and hydraulic lines running to the tail; that was good, but those independent lines ran right next to each other. A problem that knocked out one could cut the others. Further, cables and hydraulic lines all ran along the underside of the floor. McCormick pondered the fact that the DC-10, unlike the other jets he had flown up till now, offered no manual backup system if the hydraulics failed. He didn't like the prospect.

To prepare McCormick for the changeover, American Airlines summoned him to its Fort Worth training center in March 1972. One afternoon, after the formal simulator session ended, McCormick told the American Airlines instructor that he worried about losing the hydraulic system. He asked if he could stay on the simulator for some extra time and see if it was possible to control and land the giant airplane without any hydraulic system, using nothing but the engine throttles. McCormick was pleasantly surprised to find that he could. Within a few hours' simulator practice McCormick was able to take off, fly around, and land using nothing but the three throttle levers. He steered left by pulling back on the left-hand engine and advancing the right-hand engine; he steered right by doing the opposite. He made the DC-10 climb and descend by adjusting power on the tail-mounted engine; its location on the fin gave him considerable leverage.

McCormick's intuition and long experience was preparing him for a problem, foreshadowed by an obscure mishap that occurred during the initial testing of the DC-10 two years earlier.

WHATEVER IT TAKES

Jamais arrière, "never behind," was the family motto of Douglas Aircraft founder Donald Douglas. Douglas Aircraft had earned an outstanding record with its DC series of transports, most famously the DC-3. But by the mid-1960s, when it began planning the DC-10, Douglas was troubled by the memory of a sales race lost by his first jet-powered airliner, the DC-8. Boeing had gotten one hop ahead with its strategic bomber and aerial tanker contracts, and its 707 beat the DC-8 to the market and was well ahead in advance sales by 1957. In the end Boeing sold twice as many 707s as Douglas sold DC-8s. Donald Douglas never forgot, and the "never behind" attitude emerged again in 1967 when successor McDonnell Douglas began worrying about the DC-10's race with the Lockheed L-1011 TriStar. Two years later, McDonnell Douglas began getting troublesome predictions from its subcontractor General Dynamics about the DC-10's cargo door.

The problem involved a predicted chain reaction of mechanical failure that would start if something went wrong with the closing mechanism of the big outward-swinging cargo door on the rear left side of the DC-10's fuselage (see Figure 6).

Because the interiors of all airliners are pressurized while flying in the thin air of high altitude, and because the higher air pressure inside pushes with great force toward the lower pressure outside, any outward-swinging door will pop open unless it is gripped shut with some kind of latch. Inward-opening "plug" doors are safer because the interior pressure forces them tight against the doorframe, but a plug door wastes revenue-producing cargo space and such doors are heavy. Outward-opening cargo doors are standard equipment on airliners and can be safe when designed correctly. The original DC-10 cargo door mechanism required the baggage handler to do three things to close it: he had to pull down the top-hinging door and shut it, then swing down a lever on the outside of the door, and then press and hold a button that operated an electric motor at the top of the door. Putting his ear against the fuselage, he was supposed to hold the button down until he heard a click, then wait seven more seconds until he heard the motor stop running.

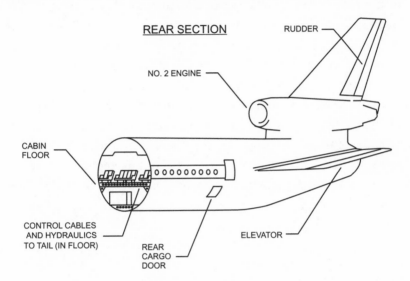

REAR SECTION   RUDDER

NO. 2 ENGINE

CABIN FLOOR

CONTROL CABLES AND HYDRAULICS TO TAIL (IN FLOOR)

REAR CARGO DOOR

ELEVATOR

PROBABLE SEQUENCE

1. Cargo handler in Detroit forces rear cargo door in attempt to shut it.

2. Electronic alarm fails to notify pilot and copilot that cargo door is not securely locked.

3. Air pressure outside fuselage drops with climb-out until pressure on door causes pins to shear. Cargo door blows out.

4. Higher-pressure air trapped in cabin collapses cabin floor near cargo door, jamming hydraulic lines and cables to tail (for No. 2 engine, elevator, rudder, and trim tabs).

Adapted from NTSB

On the inward side of the door, that motor caused latches to reach out and grasp a metal fitting. So what's the problem? If for any reason the electric motor did not finish lowering the latches all the way, the door would appear to be closed until the airplane reached enough altitude to let the pressure differential blow the door out. This false-closing problem was more likely to happen with an electrically driven door closer than with a hydraulic drive. The design was originally going to use hydraulics, but under pressure from its client American Airlines to simplify and lighten the DC-10 equipment, McDonnell Douglas shifted to an electric door closer instead. This worried engineers working for the builder of the door assembly, Convair Division of General Dynamics. Convair engineers even sent McDonnell Douglas a formal document, called a "failure modes effects analysis," describing the problem and the disastrous consequences. Convair's analysis didn't get to the FAA.

FROM THEORY TO REALITY

On May 29, 1970, McDonnell Douglas managers learned that a door blowout was not merely theoretical. On that day, during a cabin pressure test of its first airplane at a hangar in Long Beach, an improperly closed cargo door burst open. This caused the floor of the passenger compartment (holding pressurized air) to crash down into the cargo compartment (holding now-unpressurized air). But the only thing the May 1970 failure proved to McDonnell Douglas was that some joker out on an airport tarmac might not press the electric button long enough to finish the latching process. Its solution was to blame Convair for a weak floor and try to bully that company into paying for a quick fix: namely, to put a hole in the door for a vent flap that would be closed by the same linkage that shut the cargo door. The new vent flap would supposedly keep the airliner from holding air pressure unless the cargo door was safely latched. If the vent didn't shut, the pilots would know from the air leakage that there was a problem, before they got high enough for the cargo door to blow out. If the vent flap shut, the airplane was safe to fly. It was foolproof, except for one thing that nobody noticed: a little excessive force by a baggage handler, struggling to shut the door, could make the vent flap close even though the cargo door was not fully locked. Thus, the pilots would start the flight without being aware of the problem.

So it happened that on June 12, 1972, just two months after his simulator training, and with less than one hundred hours on the DC-10, McCormick got a chance to try out his new skill on a real airplane with real passengers. He was the captain of American Flight 96, originating in Los Angeles and terminating in New York, with stops along the way. During the brief layover at Detroit's Metro Airport, cargo handler William Eggert had trouble closing the rear cargo door. By leaning his knee on the closing lever Eggert got the cargo door to shut, but the little vent flap looked askew. Eggert called a mechanic, they opened and shut the door again, and decided it was good enough. A warning light in the cockpit blinked out, telling the crew the door was locked, not giving any sign that the latch on the outside of the door had signaled a "closed" position only because Eggert's weight on the latch handle had bent a metal linkage on the inside of the door. It was no fault of his. Employees by now regarded the DC-10 cargo door as difficult to shut even under good conditions, and there was no way for Eggert to know that the latches were unsafe.

Climbing on autopilot through an altitude of twelve thousand feet with sixty-seven people aboard, Flight 96 was near Windsor, Ontario, when things went crazy on the flight deck. The crew members heard a bang from the rear of the plane, and a jolt sent copilot Paige Whitney and McCormick slamming back in their seats. The left rudder pedal jammed to the floor, and the engine throttles flew back to idle. McCormick's right leg came up, and his knee hit him in the chest as a blast of dust, grit, and rivets blew into his face, knocking his headset off. The emergency trim handle broke off in his hands.

McCormick tried moving the control column back in order to level out the airplane, but the elevator controls were so damaged that he could budge the column only with great difficulty. The airplane went into a right-hand turn and began nosing into a dive that if not stopped would be its last. Cockpit warning lamps flared up from one side of the panel to the other, telling of an engine fire and dangerously low airspeed, among many other problems. The only thing McCormick and Whitney could think of as the cause was either a midair collision or a bomb explosion.

The round cocktail bar at the rear of the airplane collapsed into a crater that appeared in the floor by the burst cargo door. Flight attendant Bea Copeland fell into the pit and looked through the hole in the fuselage at her feet to see the landscape below. The air pressure had generated so much leverage on the linkages that it sheared off metal pins and blew the door out. The rear cargo door broke in two, folding the top part up like the lid of a tin can and sending the bottom to crash against the tail and fall to the ground. Just as happened in the 1970 hangar test, the cabin floor near the door collapsed. The floor collapse jammed the control cables to the tail.

McCormick pushed the wing-engine throttles to full power, bringing the airplane out of its dive. To counteract the right-hand bank from the jammed rudder, he turned the wheel on his control column forty-five degrees to the left and kept it there. Believing that he who hesitates is usually better off, he paused to take stock of the situation.

The airplane was still in a very dangerous state, so what McCormick had accomplished at this point may not sound like much. But he had squeezed out a tiny margin of safety, a few minutes to think and to talk things over with his copilot and engineer before committing himself. The significance of winning little safety margins like this on the machine frontier can't be overstated. It's akin to the way that escape artists prepare to break free of ropes being knotted tightly around their arms and legs: by arranging their limbs just so and flexing their muscles, they can steal themselves enough slack to get loose later.

McCormick's simulator training had taught him to avoid sudden moves, because the tiny edge of control he still had could not extract the plane out of a steep dive or turn. He and Whitney alerted air traffic control and shut down the tail engine by starving it of fuel. McCormick nudged the wing engine throttles to see if he could control the airplane in the same way he had in the simulator. It worked; the DC-10 had a fatally flawed cargo door design, but, as McCormick knew, the layout of its engines made it unusually well suited to steering by engine power.

Turning the wheel on the control column to the left leveled the airplane's wings but left him without enough use of the ailerons to control the airplane's path. In one respect he was a little better off than he had prepared for in the simulator sessions—he had some use of the elevator control, though only one side was working and every time he used the elevator it tried to roll the airplane over. But in another respect he was worse off than in his simulator training because he had no control over the tail engine.

At this point the emergency procedures manual called for an emergency descent to lower altitudes, but McCormick overruled that. The passengers could survive the thin air of twelve thousand feet for a few moments, but his control over the airplane would not survive any sudden moves. He could probably keep the airplane under control in midair, but could he steer it back to a runway?

Flight attendant Copeland pulled herself to safety from the collapsed floor, and McCormick coolly informed the passengers via the intercom system that American Airlines would provide a new plane at Detroit for them to continue their trip, because there was a "mechanical problem." The passengers' mood lifted immediately: suddenly they could see beyond this apparently fatal problem to life on the ground.

Working very slowly, McCormick turned the jet back to Wayne County Airport. In a feat of airliner piloting that has to date never been equaled, and on his first try at landing, McCormick kept the crippled DC-10 under control to the runway threshold. It came in hot, at 186 miles per hour, because the only way McCormick could keep the airliner from tipping forward and smashing into the ground was to maintain high power on the engines. Immediately after the airplane touched down, the jammed rudder sent the airliner off the runway to the right, the nose gear threatening to break off every time the airplane slammed across a taxiway.

Now, irony of ironies, the airplane was speeding toward a crash with the airport fire station. Both wing engines were at full reverse but they weren't stopping the aircraft. Copilot Whitney seized the moment and steered the DC-10 to safety by cutting back on one thrust reverser while adding power on the other. This overpowered the stuck rudder. The wheels finally came to rest half on concrete and half on grass. Not a person was killed, and this aircraft would fly again, most recently for Federal Express.

McCormick asked McDonnell Douglas to "fix the damn door"; but in 1974 another DC-10 lost a rear cargo door due to a faulty latch mechanism because McDonnell Douglas never installed the promised reinforcement package on that aircraft at its Long Beach factory, though somebody there had wrongly stamped the inspection paperwork with a quality-control approval, signifying that the work had been done. And worse luck, unlike Captain McCormick, the Turkish Airlines pilots on the 1974 flight had not been prepared for such an emergency. To boot, their airplane was fully loaded and therefore would be harder for even a master like McCormick to control. The Turkish flight crashed in a French forest on a flight from Paris to London, killing 346.

GET OUT

Airplanes can be almost as dangerous sitting on the ground as they are in a crash landing, if a fire is involved. Every once in a while a cabin fire puts a hundred people or more in a very desperate situation. It would be good for you to think about this particular demon ahead of time, because you won't find out what to do from the preflight safety talk, and if it happens on your plane you won't have the time or mental clarity to think it out. It happens when a jet's fuel supply erupts into flame, possibly because of an engine explosion or a low-speed crash on the runway. Suddenly that tidy fuselage you walked into is a roaring, smoke-filled deathtrap from which you and everyone else who is still alive have few minutes, possibly only seconds, to escape.

To get some idea of how hard it would be to cope on such a day, imagine having to take the most difficult final exam of your life while somebody is lobbing tear-gas grenades at you, when you are also suffering a major migraine headache and violent food poisoning.

Early in the morning on August 22, 1985, a 737 jetliner began its takeoff roll at the Manchester, England, airport with 131 passengers, mostly tourists, on board. About halfway along runway 24, the combustion chamber on the left-hand engine cracked and flew apart, throwing a hunk of metal through an access panel on the underside of the wing. The captain heard the thump of impact. Thinking it was something minor such as a burst tire, he had the copilot apply reverse thrust on the engines to slow the plane but told him not to bear too hard on the brakes. Then they steered the jet onto an access road to clear the runway for other airplanes to take off. It was a considerate gesture—Manchester had only one active runway—but it meant an extra twenty seconds of delay.

Meanwhile the passengers at the back saw a bright glow on the left side as burning fuel streamed out a plate-size hole in the wing tank. Worse, the airplane had stopped at an angle that allowed a breeze to push flame against the fuselage. Smoke seeped through windows on the left side, and in short order fire burned its way into the tail, breaking out windows. Now a black greasy smoke boiled along the cabin roof toward the nose. Some passengers hustled their way up the aisle; some fumbled for carry-on luggage; others crawled over the tops of the seat backs.

Perhaps half a minute after the smoke first broke into the cabin, it pushed down from the ceiling to head height, deadening all loud noises. There was no evading this smoke short of reaching an exit door. People in a house fire usually have the chance to crawl along the floor and gain a minute or two before smoke fills the rooms, but not a hundred desperate people jammed in a narrow aisle. To drop to the floor was to be trampled; simply to reach for a handkerchief to cover the nose and mouth was to risk being knocked off balance. The pressure was jamming people into the narrow exit doors. Keep in mind that this crowd behaved exceptionally well, without the frenzied shoving that usually occurs during such crises.

The poisoning effect of inhaling the smoke of burning plastic and jet fuel was instantaneous. The first breath was so horrendous the body reacted by taking a second, deeper, breath, something like the involuntary gasp one might make when jumping into cold water. Eyes, nose, and throat filled with blobs of black foamy tar. Muscle strength failed, and passengers lapsed into a stupor from the fumes' effects. People tripped over each other as their range of vision fell to a few inches.

Fifty-five people died on the airplane that day, most of them succumbing to smoke inhalation rather than heat. Cyanide and carbon monoxide were the most dangerous chemicals among the hundreds in the black and sooty brew. And the lessons? Some seasoned passengers who have heard about such fires have begun to carry "smoke hoods," which can shield the lungs and eyes from smoke. Regardless of whether you decide to buy one, there are other precautions you should be aware of. During the safety talk, while everyone else is reading the newspaper, count the seat backs between you and the nearest exit. In case of fire, you'll need to know this because the smoke will be so black and thick you won't be able to see the exit door from the aisle. Forget trying to stay low, below the smoke; you'd get trampled unless the airplane is nearly empty. Disregard whatever door you walked through when boarding from the airport, unless it happens to be the closest safe one to you. Studies of fuselage fires and breakups show that most people try to fight their way down the full length of the cabin to that same door, walking right by open exits. If the airplane is stopped and you know it's on fire, get out immediately. In 1957, passengers stood around or died in their seats in a DC-6 fire in New York, a few feet away from exit doors they could have opened easily.

As recently as 1995, when Boeing was taking its new 777 airliner through certification, the FAA made the company hire volunteers to try out the cabin evacuation plans during a simulated emergency. Boeing satisfied this requirement by parking one plane in a darkened hangar, turning out the cabin lights, and asking flight attendants to hurry people to the emergency exits. Now the FAA has stopped requiring live testing, in part because the drills were too dangerous. People were slipping off the escape slides, a few even breaking arms and legs in falls. Others fell to the floor in the simulated rush.

But there is something to be said for live training that is realistic enough to see what havoc the really bad day might bring. Think of the men of the *Ocean Ranger*, when they tried to use their lifeboats for the first time in a storm. The boats had worked fine during the drills on afternoons in calm weather. A boat simply descended sixty feet to the water on its long "fall" lines, and the men on board started the little engine, released the falls, and motored away from the rig. So it must have infuriated the men of the *Ranger* to see how a storm and a capsizing rig altered the plan. Suddenly the craft weren't lifeboats at all. Battered by waves and smashed against the rig's structure, flipping over when the men tried to climb out too fast, they were death traps.

Emergency training needs constant renewal. Walter Stromquist, who as a beginning mathematician rode as an observer on submarines as part of a two-year risk-research program for the U.S. Navy's Development Squadron 12, recalls the routine on the boats he rode: "Every day, all day, they are studying or taking exams, challenging each other on maintenance issues, and running emergency drills. There was at least one simulated fire and one 'emergency' reactor shutdown every day." In one improvised exercise, a submarine commander told a sailor at a control panel to "play dead." The commander then changed the setting on a critical piece of equipment and clocked the time until someone came to diagnose and fix the problem.

FANTASIES TO KEEP US COMPANY

Realistic emergency training may strike you simply as a way to imprint certain skills on people, but there's another benefit: breaking through the misplaced self-confidence that too many of us carry around. David Dunning and Justin Kruger, psychologists at Cornell and the University of Illinois, wanted to dig further into the known fact that taken as a whole across large populations, the average person rates himself as above average in skills and knowledge. Plainly, the average person cannot be above average in his or her skills, so how is it that we do not know ourselves? Ignorance breeds overconfidence, they say, particularly in fields in which people know just enough to be dangerous. When Dunning and Kruger surveyed and tested Cornell undergraduates on such skills as logical reasoning and humor, the same pattern was repeated every time. Those with the worst test scores grossly overrated their performance and skills compared to others. The two psychologists quote Charles Darwin on this phenomenon: "Ignorance more frequently begets confidence than does knowledge."

The two have a few theories about how this could be: maybe incompetent people have trouble observing the world around them, or maybe the world does not provide good feedback. "It's possible that as humans we desire to be ignorant of our abilities," Kruger says. The two researchers did find that when they picked out the poor performers and taught them the skill the students thought they had, but actually lacked, it made them much more accurate in estimating their own abilities.

We can harbor our fantasies for a long time, even a full lifetime, as long as nothing occurs to upset them. After I earned my private pilot's license I daydreamed occasionally that if called upon while on an airline flight, I could probably land an airliner in an emergency. Ten years later I had the chance to fly a Boeing 737 simulator at Delta Airlines' headquarters in Atlanta. Captain Jim Kater briefed me on the controls and began by flying the tricky approach along the Potomac River to Reagan Airport in Washington, D.C. Then he moved the simulated airplane to the Atlanta airport. I pushed the throttles forward and got into the air in one piece, but I found myself all over the sky. Flying the darn thing was not like I expected at all. I swerved, dived, and climbed. The control column took real muscle to budge, and it needed constant adjustment from a trim knob to make it manageable. The airplane lagged in responding, so I was always behind it and hence overcorrecting. Kater pointed me back to the approach for landing. A mechanical voice yelled "Pull up! Pull up!" and "Glide slope" at me. I managed to control it well enough to crash my load of simulated passengers a mile from the airport. On the second try I touched down on the runway, but it was 90 percent due to the angel at my right elbow.

I was seeing one of the hidden virtues of harsh training on a realistic simulator. It teaches failure to the Walter Mitty crowd: failure so vivid and complete that it breaks through unjustified confidence that we can wing our way through a full-stops crisis. After receiving a bucket of cold water like that, we can start learning how to handle some really bad days.

Even without a multi-million-dollar simulator, it could be a lifesaver simply to sit down and spend a quiet hour thinking hard and skeptically about how equipment is supposed to work in a real emergency, with smoke, fire, storm, tilting decks, and much anxiety. I think of the French warship *Liberté*. Bags of "Poudre B" smokeless gunpowder self-ignited in a forward magazine while the battleship was at harbor in Toulon, France, on September 29, 1911. Dense yellow smoke curled up from the forward ammunition magazines as the commanding officer tried to mobilize damage-control teams at the stern.

There was not much time to act. In the bow, heat was softening steel racks, and soon hundreds of rounds of cannon ammunition would be falling into the fire. The ship's commander sent the chief engineer and chief gunner to the bow to fill the forward magazine with seawater. The magazine lay below the waterline, and the two men were supposed to open seawater valves to flood the space and put out the fires. The engineer and gunner in trying to reach the seawater valves found their way blocked by fire, smoke, and explosions.

The flooding valves for the magazine were located below decks but directly above the ammunition—which was on fire and prevented the men from approaching the valves. Twice the men risked their lives to reach the valves, but they had to retreat each time. Then the ship's electricity went out, putting all the lower areas in darkness. The men reported to the commanding officer that it was impossible; he abused them and ordered them back to finish the job at any cost. The engineer and gunner went back, never to be seen again because the whole front third of the *Liberté* exploded soon afterward.

In my law school, this catastrophe would have been called "foreseeable and avoidable" because three French warships had already suffered explosions and fires after the spontaneous ignition of Poudre B. Poudre B had totally destroyed the battleship *Iéna*. By the year of the *Liberté* blast, safer competing gunpowders were widely available. Poudre B, invented by French chemist Paul Veille in 1885, was made by blending nitrocellulose with ether and amyl alcohol. With time the stored pellets degraded and released that alcohol once more, which mixed with nitrogen to make dangerous amyl nitrate and amyl nitrite compounds.

OUR FINELY CALCULATED RISKS

Which contingencies need a plan, and which are too freakish to worry about? The official methods for figuring the risks include failure mode effects analysis, fault trees, and event trees. These techniques attempt to sort through the most pressing probabilities. But there's always a way to tilt the figures, and the history of these methods shows some attempts over the years to label worrisome events as extremely remote as a way of expediting a particular project; it happened before the DC-10 cargo-door blowouts and the *Exxon Valdez* oil spill.

Before the *Challenger* disaster, NASA used a "probabilistic risk analysis" (PRA) method to report the odds of a disaster on any given shuttle mission. There were several outcomes depending on technique, but the most commonly cited one was one failure in one hundred thousand flights. A PRA began with hypothesizing about sequences of mishaps that could lead to a specific critical failure, then used engineering estimates to assign a probability of occurrence for each of those mishaps. After the first few flights demonstrated that the reusable heat shielding would function if repaired religiously after each flight, most of the engineers' concerns came to rest on potential problems with the three main rocket engines or their fuel.

Because as humans we focus our attention by concentrating only on what we regard as likely to occur, such opinions can literally affect how we see the world. In 1981, I was interviewing a professor at Texas A&M for an article. I took a chair facing him across his desk and made notes as he talked. About fifteen minutes into the interview his waste can, located behind him and out of my sight, burst into flames that were at least four feet high. We both jumped up; he threw some kind of a cover over it and quickly snuffed out the flames. The fire had been caused by a cigarette butt some other visitor had dropped in, he guessed.

As he put the fire out I remembered seeing a quick flash of flame behind his chair maybe half a minute before we became alarmed. Call it a flashback: just below the level of awareness, I had glimpsed the flash behind his chair while I was taking notes. It had sprung up for a split-second and then disappeared from sight. Whatever passes for the current-events center of my brain had noted the event but immediately set it aside as so improbable a sight that it need not be dealt with by the conscious mind.

Most of us sort out our probabilities not with statistics but according to what we have actually experienced. Psychologists call this the application of "heuristics." Heuristics are the generalizations we accept about the working of the world, drawn from the grab bag of what we've seen personally or what we've heard about from trusted sources like friends and relatives. Because heuristics seem to be based on the most credible of all information—the evidence of our own eyes and the word of those whom we trust—people place great value on those beliefs. A person's heuristics do shift with time, as other information crowds out the old or after once-vivid memories fade. Heuristics explain why after a well-publicized flood so many people rush out to buy flood insurance, then let it lapse before the next deluge.

Heuristics underlie how we regard risk in our daily lives. My kitchen illustrates: according to something my wife heard, food left in open metal cans in the fridge will spoil, so she doesn't ever store food that way. A friend of my family burned his hand as a child when he accidentally yanked something hot off the stove, so I always turn pot handles toward the wall when cooking. And after refilling a butane barbecue lighter the whole thing blazed up in my hand when I next tried to use it, so I haven't bought one since.

Heuristics are at the root of many superstitions. For example, first we go about our daily routines a certain way and things happen to go well; then one day things go badly and we look back on what we did earlier that day to determine what we did that was different from our established pattern. Whatever was different must be the reason for ill fortune later in the day, so we try to avoid that circumstance in the future. I was surprised to find that some otherwise hardheaded engineers at the Kennedy Space Center would sooner forgo watching a shuttle launch than come to Kennedy on launch day without their lucky pieces of clothing on. While riding on the shuttle booster recovery ship *Liberty Star*, I heard of one technician who still grieves because he wasn't wearing his lucky shirt when the *Challenger* went down.

FIVE THOUSAND TONS OF TROUBLE

One heuristic principle says if nothing very bad has happened to a system or factory so far, it probably won't happen in the future, either, whatever the chemistry involved. It helps explain how it came to be that a plant making ammonium perchlorate, which is a powerful oxidizer used in solid rocket fuel, happened to be in a built-up area of Henderson, Nevada, when a fire started there on May 4, 1988. Ammonium perchlorate (AP) is the oxidizer that goes into solid fuel for civilian space boosters and rocket motors for battlefield and intercontinental missiles. The factory, which began operating in 1958, was run by Pacific Engineering & Production Company of Nevada, or PEPCON. Henderson was also host for the nation's other principal maker of AP, Kerr-McGee.

The fire department's representative said afterward that the disaster began after a welder's cutting torch ignited AP stored at the plant. PEPCON and its consultant, Exponent Inc., instead traced the root cause to natural gas percolating through AP being stored in a bin, saying the gas came from a leaky sixteen-inch transmission pipeline. The plant was storing five thousand tons of AP at the time, more than usual because of the halt in space shuttle flights as a result of the *Challenger* disaster.

Whatever its cause, the fire spread so quickly that just minutes later the company firefighters abandoned their attempt to stop it. The plant had seventy-four employees on-site and most reached the exits in time. Ten minutes after the first flames, an explosion raised a mushroom cloud over the plant. Three more blasts followed, each growing in intensity, making buildings sway in downtown Las Vegas, twelve miles away. The explosions killed two PEPCON executives at the plant, injured more than three hundred people, tore doors off hinges two miles away, and burned down the Kidd & Company marshmallow factory. Altogether, the fires and explosions consumed about forty-five hundred tons of ammonium perchlorate. Insurers estimated damages of $75 million.

Dan Evans, then director of the Nevada Occupational Safety and Health Administration, said afterward that people assumed that since

PEPCON had been making rocket fuel ingredients for twenty-five years and nothing had happened, nothing was going to happen in the future, either. "Of course," he said, "that's when all hell breaks loose."

Experienced workers in potentially hazardous situations usually have a few danger scenarios at the forefront of their minds, as part of their workplace heuristics. Call them trade fears. Airline pilots worry about collisions with private airplanes blundering across controlled airspace. People who make nitroglycerine watch out for overheating during manufacture, and particularly any sign of red fumes. Experienced tower workers don't worry about slipping off—it rarely happens, given the safety equipment they use—but they do worry about a mishap while moving the "gin pole," a temporary derrick bolted to the top of a broadcast tower that's under construction or maintenance. A massive gin pole swinging out of control can fatally weaken the whole delicate structure and bring everyone on it crashing down. In 1998 some kind of problem with a gin pole was the cause of a collapse of a 1,550-foot-tall broadcast tower in Cedar Hill, Texas, which in some freak of physics hurled a worker a third of a mile.

Those people who write about industrial catastrophes hear the criticism from time to time that such books are only Monday-morning quarterbacking, loaded with twenty-twenty hindsight. Would that all catastrophes left such clear records: in some cases the string of failures came so thick and fast, almost simultaneously, that no one could unravel the exact sequence afterward. And in some cases, such as tall tower collapses, Payne Stewart's Learjet crash, or the PEPCON blast, the final cataclysm was so savage that the critical evidence got shredded, flattened, or pulverized. People will never agree on the first failure, the trigger, in such cases. Still, a short list of contributing causes usually emerges from the rubble.

Watching out for signs of known problems is good, but as systems get bigger and more complex we have to remember that our Achilles has many heels, so to speak. Some of the problems that arise will never have come up before, so the simple hindsight of heuristics can't save us. We can hope that workers don't anticipate danger from just one direction, since some of the worst disasters arose out of problems so minor that people had previously discounted them, as with the *Challenger* disaster.

We've seen how tempting it has been for organizations to set disaster precautions aside as too expensive, too embarrassing, or an interference with production. The complete lack of preparation didn't worry the managers on the scene beforehand, because they had convinced themselves that any such terrible scenario wouldn't happen for years, if ever. But as the next chapter shows, managers who look outside their familiar surroundings may discover that calamity is not going to wait that long.

6: TUNNEL VISION

GO AWAY, I'M BUSY

On the morning of April 10, 1963, the nuclear attack submarine *Thresher* began a day of diving tests two hundred miles off Cape Cod. The water was eighty-four hundred feet deep, which was plenty of sea room for running the boat to its test depth of about thirteen hundred feet. The captain notified the escort tug *Skylark* of the plans for the *Thresher* and its 129 men. After an hour the *Thresher* sent the message via underwater telephone that it was heading to test depth. The *Thresher* descended in a spiral. Fifteen minutes later, another call came up: the boat was experiencing minor difficulties and attempting to blow ballast and surface.

According to the naval court of inquiry afterward, a seawater-piping joint brazed with a silver alloy had probably burst, thus opening up the engine room to the sea. Submariners are keen on tradition, so it's likely that somebody down there recalled the circumstances of a seawater leak on board the first *Thresher*, a diesel-powered boat that cruised the Pacific during World War II, hunting Japanese ships. Seawater came close to shorting out that boat's electrical gear on July 5, 1942, after a patrol boat forced her to crash-dive near Maloelap Atoll. At a depth of 250 feet, water began leaking through the outer hatch and cascading into the control room. The cause was determined to be a leather sandal, which had come off as a sailor jumped down from the bridge. The sandal had caught between the hatch and its frame, keeping the hatch from closing. Unable to stay down and unwilling to surface under the gunboat's artillery fire, the skipper ordered the boat to rise to a level barely poking above the surface. This gave the first *Thresher*'s crew just enough time to clear the hatch and dive again before the Japanese guns could hit the submarine.

That wartime scrape had been bad enough, but a leak opening beyond one thousand feet below the surface made it very difficult for the second *Thresher*'s crew to get back to the surface at all. Seawater from the leak sprayed across an electrical panel, and the single S5W reactor automatically shut down, cutting off the main source of power. The crew couldn't restart the reactor in time, nor could they empty the ballast tanks despite having set the ballast controls to "full blow." Air rushed out at full volume from the compressed air storage tanks, but then it choked off as ice formed from a Venturi effect as air pushed through a metal debris strainer. The ballast-blow problems lasted long enough for the submarine to plunge into serious trouble. The batteries on board could not drive the burdened vessel to the surface. The *Thresher* turned to the seafloor, dragged down by the seawater flooding in.

The submarine broke apart on the way down. The navy pulled up as many pieces as it could from the seafloor and took the lessons of the second *Thresher* very seriously. It set up a "Subsafe" program that improved construction quality, discarded all brazed connections in seawater piping, and improved emergency training. The navy redesigned its submarines for fast and foolproof deballasting, giving the sailors a pair of "emergency blow" handles on top of the ballast control panel. To guard against fatal power loss from a reactor "scram," the navy provided a way for the operators to block an emergency shutdown even at the risk of equipment damage. Although seemingly limited to the matter of submarines, the findings of the *Thresher's* court of inquiry deserved attention by others along the machine frontier.

This chapter is about people so tightly focused on a goal, so consumed with the job ahead, that they refuse to heed information coming from outside their workaday world. Tunnel vision like this is going to waste people and machinery because disasters and near misses arising out of one field can provide valuable warnings to another. Sometimes the connection is not immediately obvious.

Ponder what the *Thresher's* lessons could mean to some of us, though we may never set foot in a single submarine. That disaster had been set off by water shorting out a critical electrical panel. One disaster that the tale of the *Thresher* might have helped prevent, if taken to heart by the right person at the design stage, was the sinking of the *Ocean Ranger* nineteen years later. The *Ranger's* capsize and sinking was also triggered by seawater shorting out an electrical panel, splashing over the ballast controls after a window broke out. During early over-ocean tests of the 747, a pilot's cup of coffee overturned and shorted out the inertial navigation electronics. It would seem that any manager or designer of a complex system should know by now the importance of keeping liquids out of electrical gear.

Looking even further out of the tunnel, all systems operators should be generally aware of the havoc caused by electrical problems beyond water that bridges across circuits. Industrial insurer Hartford Steam Boiler ranks electrical problems as the number one source of its damage claims. To see how such an electrical problem can cause so much damage, consider a small short circuit that happened on November 24, 1997, at the BHP Port Kembla Steelworks on the coast of New South Wales, Australia. The fault was in a transformer providing power to cooling pumps. A fault protection device that was supposed to clear the problem didn't work, and the excessive power made a high-voltage cable overheat at the point in which it passed through a cable tray. The overheating cable lit fires at several points, and these triggered all five incoming thirty-three-thousand-volt power lines to shut off. The failing power shut down the steelmaking furnaces, which

sent fuel gas shooting seventy feet high out of the tops of the coke
ovens, igniting simultaneous fireballs across the plant's superstructure in
a scene that looked like something out of the Wizard of Oz's throne
room. Three thousand workers fled the plant. According to the worker's
union (the company denies it), the power failure also caused 150 tons
of molten steel to gush out of the number 4 slab caster. In the union
account, the steel flowed across the factory floor, set fire to two vehicles, and poured down a stairway. It was a classic industrial chain-reaction event: no fatalities, workers back in a few hours, but millions of
dollars in damages.

MAN, MOON, DECADE

The bigger and more urgent the project, the more likely it is that people will miss problems lying just outside their tunnels. One of those
problems played into the *Apollo 1* spacecraft fire on January 27, 1967.
On that day, during a test of the electrical system and pressurized to
16.7 pounds per square inch of pure oxygen, something ignited a fire
while the three-man crew was sealed inside. Plastics in the capsule
burned so fiercely that the gas pressure blew out a wall of the capsule.
The astronauts couldn't get out in time and died of smoke inhalation,
their last recorded cries for help coming just twelve seconds after commander Gus Grissom's first mention of fire in the cockpit.

The *Apollo 1* disaster was yet another example of people so devoted
to a goal, so feverish in their daily activity, so narrow in their vision that
lessons from mishaps and near misses outside their own program made little impression. But long before Apollo, high-oxygen atmospheres had
been causing deaths and injuries of a strange and horrifying nature.

One April day in 1996, while spending a day with my son in the
hospital, I saw a label glued on the wall near the oxygen-supply outlet.
"Use No Oil," it said. That was thoughtful: I guessed that without the
label, workers using WD-40 somewhere along the line might cause a
little oil to get into people's lungs via the oxygen hoses. Maybe oil

would cause an allergic reaction. As it turns out, this is not at all the reason for the label.

If you walked up to people at the airport with a clipboard and a friendly expression and asked them if they thought oxygen is a good thing, chances are that almost anyone who took you seriously enough to respond would say something like, "Yes, I am all for oxygen." And they'd be quite correct because what they would be approving of is oxygen as we know it, that is, holding steady at 21 percent of the atmosphere by weight. Oxygen is odorless, invisible, and highly esteemed in action-adventure movies about spacecraft and submarines. Out in the real world, American businesses use twenty million tons of oxygen a year for steel-making and chemical manufacturing. It saves lives and promotes healing. It launches our rockets. Because we hardly ever encounter such conditions, it wouldn't be quite fair for you to persist, "Yes, but what if oxygen was 35 percent of the air? What if oxygen was 100 percent of the air?"

Oxygen has a key role in some famous disasters and near misses—*Apollo 1, Apollo 13*, and the ValuJet airliner crash to name a few—for a good reason. As the amount of oxygen in a room goes up, garden-variety materials take on a fearsome aspect. Years ago very few people stored oxygen in their households, and if they did so it was in a tank in the garage, for metal cutting or welding. Now hundreds of thousands of households have oxygen handy, in the form of home oxygen delivery kits.

These rely on Thermos-style tanks storing up to seventy-five pounds of liquid oxygen. That's more than enough to give a room a very high ratio of oxygen if something on the kit fails and starts venting gas. In such an atmosphere even a static spark can set a carpet on fire, and as more oxygen gushes out it can lead to an uncontrollable fire the likes of which firefighters have come to dread. Even their sturdy Nomex suits will burn up in such an atmosphere.

Oxygen is an extremely common element, when you include its chemical compounds. The average rock you run across is about half oxygen by weight, and water is nearly all oxygen by weight. Oxygen by itself can't burn, but as its amount increases many substances turn into a rich fuel. Take an iron pipe, put a bundle of welding rods inside it, and run oxygen through the pipe; once the end is lit with a cutting torch, this simple device becomes a "burning bar," a device of fiery mien that will burn through concrete, diamond, the reentry shield of a spacecraft, or anything else in its way.

A long series of tragic fires in hospitals and EMS vehicles teach that combustibles and pure oxygen do not mix. Sometimes plastic masks have caught fire and caused painful skin burns. In the most tragic cases, patients found themselves inhaling a jet of flame. Even metal filings in a metal pressure line, tumbling along with the flow of gas, can get hot enough with friction to start a fire inside the line. In 1987 a sparking toy gun ignited a fire inside a "hyperbaric" pressurized-oxygen chamber used for medical therapy, killing the boy inside. Ten years later another fire in a hyperbaric chamber in a hospital in Italy killed ten patients and a nurse.

Massachusetts firefighter Jon Jones, then an analyst for the National Fire Protection Association, interviewed an aircraft mechanic after the January 1987 crash of a Bell 206 medevac helicopter in North Carolina. The helicopter had caught fire in midair and gone down, killing everyone on board. Jones helped trace the origin to a dab of oil wiped onto an oxygen fitting before the flight.

"As I was talking about the cause, the mechanic got almost apoplectic," Jones says. "I asked him why, and he said nobody had ever told him about this." Jones knew then how avoidable the helicopter in-flight fire was. While serving in the air force in the mid-1960s as a policeman guarding jets on the ramp, Jones had often gotten reminders about the danger of bringing petroleum products near the planes' life-support systems. Clean isn't enough for high-oxygen atmospheres, said the military; it has to be "LOX [liquid oxygen] clean," meaning devoid of combustibles like oily films.

Indeed, to a wide range of oxygen users, and well before the *Apollo 1* fire, the danger of high-oxygen atmospheres was a known fact. As Jones said, the air force maintenance people knew. And after a fire in an oxygen chamber in March 1961, those involved in the Russian space program knew. At the end of ten days, cosmonaut trainee Valentin Bondarenko removed sensor pads from his skin and inadvertently threw an alcohol-soaked pad of cotton onto a hot plate he used for cooking. Supercharged by the oxygen, the cotton pad blazed up and ignited his suit. Bondarenko died eight hours later.

The U.S. Navy knew. At a Pearl Harbor pier on June 15, 1960, Joseph Smallwood was acting as auxiliaryman and supervising the loading of liquid oxygen from a tanker truck onto the nuclear submarine *Sargo* when something went wrong. He ordered another man out of the compartment and stayed to fight the fire. The compartment exploded, sending flames more than a hundred feet high. The captain had to flood and sink the *Sargo* at the stern to put the fire out. This fire caused the navy to change the way it loaded oxygen onto submarines.

THE BUILT-IN DEFECT

Despite the evidence building up outside the space program about the dangers of oxygen carelessly used, none of the message got through in time to prevent the mistakes made in the first Apollo-manned spacecraft. We'll call this spacecraft *Apollo 1* for simplicity, though the mission never had that designation until after the fire. To the people on the scene, the capsule was "Command Module 012."

Designing the spacecraft to guard against fire in high-oxygen atmosphere was NASA's responsibility, since that agency wrote the details of how the Apollo spacecraft was supposed to work. Two near-fatal mishaps in the space program captured NASA's attention and distracted managers from the dangers of oxygen and plastic in a tightly closed chamber.

The first near miss was a malfunction with the Mercury life support system in April 1960. Up until then NASA was using a less-flammable mixture of nitrogen and oxygen, to be kept at the low pressure of five pounds per square inch while in space. During tests of a full-scale model in a vacuum chamber, nitrogen gas seeped into the spacesuit oxygen supply and almost asphyxiated G. B. North, a test pilot in the capsule. Lesson learned: give the astronauts pure oxygen to breathe instead, both when waiting on the ground prior to launch and in space during flight.

The second mishap was Gus Grissom's first flight as a Mercury astronaut in July 1961. After his one-man *Liberty Bell 7* capsule splashed down in the Atlantic Ocean, the hatch blew off before the arrival of navy frogmen, and the capsule flooded and sank. Grissom claimed the hatch blew off on its own, without him triggering the explosive bolts to the hatch, and this was NASA's official line. Others suspected that some mistake Grissom made was the cause, but there was no way to prove it. The lesson NASA took away was to avoid hatches that open outward and can be removed too easily, because a mistake in space could dump all the capsule's air. So the *Apollo 1* capsule was built in such a way that getting the astronauts out of the capsule took at least two minutes. To open the innermost door, the astronaut sitting nearest had the job of unfastening it at six places with a ratchet handle, then pulling it off the frame and into the cabin. This left another door plus an outer cover to pull off, which technicians did from the outside. After the fire, astronauts got a capsule with a one-hatch door that they could open in ten seconds.

Given that NASA had flown sixteen manned Mercury and Gemini missions all using a pure-oxygen atmosphere, without a single capsule fire, the contractor—North American Aviation—who built the capsule came under intense scrutiny afterward. In the national shame and anger, North American's proud history didn't count for much. This was the thirty-nine-year-old-company that built the tough B-25 bombers that General Jimmy Doolittle's men had flown off the aircraft carrier *Hornet* to bomb Japan in April 1942. It designed and manufactured the P-51 Mustang long-range fighter, which was so critical to winning the air war over Europe. By 1961, when it applied for a piece of the Apollo action, North American's résumé also showed two jet fighters, the X-15 rocket plane, and the Navajo and Hound Dog missiles.

North American won the right to build three major components for Apollo: the second-stage rocket booster; the command module, the conical capsule that astronauts rode in during launch and re-entry; and the service module, a cylindrical section behind the capsule that provided life support and propulsion for the long voyage between Earth orbit and Moon orbit.

NASA managers and the three astronauts went to North American's plant in Downey to decide whether to accept the capsule. They did so, and the contractor delivered the capsule for the *Apollo 1* manned flight in August 1966. By then the company was in serious trouble with the head of NASA's Office of Manned Space Flight, General Sam Phillips. In December 1965 Phillips wrote a long memo to North American president Lee Atwood saying that the company's Apollo work was gravely behind schedule and over budget, and doing badly in quality control. In a private letter to a NASA official Phillips said bluntly that he had lost confidence in the company.

North American managers held about the same opinion of NASA, laying particular blame on thousands of changes that the space agency had wanted along the way. Though North American would get most of the blame for the fire afterward, vulnerable as it was to accusations of poor workmanship and loose documentation of changes, to their credit the company and its subcontractors did warn NASA about fire danger, in writing, on several occasions and far enough ahead of time to make a difference. It couldn't have come much clearer than it did in a letter from Hilliard Paige, an executive with General Electric's space division. "The first fire in a spacecraft may well be fatal," Paige said, and warned NASA about getting overconfident just because Mercury and Gemini spacecraft had suffered no fires with pure oxygen atmospheres.

This fussing between the players was all behind the curtain; meanwhile the pace never slowed. Working around the clock except for Christmas and New Year's Day, the Kennedy Space Center launch team put the capsule through the required battery of tests. By January 6 the capsule was in place at pad 34, sitting 215 feet high on top of the three rocket stages.

At this point the Apollo program was riding high as well. Apparently no one outside the program knew of the Phillips memo. One of the celebrities as *Apollo 1*'s launch neared was the hard-driving

Apollo program manager, Joe Shea. A *Time* reporter was following Shea around to gather material for a cover story to appear on the week of *Apollo 1*'s flight.

Shea deserved the praise. Measured by his accomplishments to date, a strong argument could be made that he had earned the right to be the first man on the moon himself, as much as any astronaut. Shea learned the rocket trade by doing systems engineering on the Titan 2 missile project for a division of General Motors. He showed a rare talent for managing and motivating people while keeping up his superb engineering skills. During a corporate crisis in meeting the Titan contract, Shea began living at the plant around the clock, five days a week, to make sure he was available whenever needed. Staying overnight let him see everything that was going on, he explained. A man who worked as hard as this could have justifiably quoted the saying of Alfred Nobel. When asked where his home was, Nobel had said, "My home is where I work, and I work everywhere."

Accepting the NASA job of Apollo program manager in December 1961, Shea had swept up a milling mass of engineers and brought the moon within their reach. By 1964, NASA and its contractors had an astonishing three hundred thousand people working on the project, achieving things thought impossible a few years before. When the sheer mass of problems threatened to send his people into a panic, Joe Shea soothed them by saying that the work was not that complicated, and it would get done one piece at a time.

Shea's main management tool wasn't complicated, either, just a looseleaf notebook that his staff filled each Thursday with progress reports, crisis bulletins, and cost figures sent in by every branch of the Apollo program. Shea marked up the pages by working through the weekends, then releasing his incisive comments on the following Monday to be answered in time for the next notebook. Tyranny would come easy to people in such a high post, but Shea maintained his upbeat attitude and sense of humor throughout, at least until the fire.

On the afternoon of Friday, January 27, 1967, Gus Grissom, Ed White, and Roger Chaffee climbed into the capsule for what was supposed to be a few hours' worth of routine work. NASA considered none of the tasks hazardous. In the "plugs out" test, technicians would be checking the capsule to see how well it could shift over to internal power. The tests would also give the astronauts a chance to rehearse the notebook full of procedures for launch and orbit while wearing their pressure suits and having to rely on the suits' intercom links. Pure oxygen filled the capsule to 16.7 pounds per square inch, a little higher than air pressure at sea level.

Moments after they plugged in their suits' oxygen hoses, Grissom radioed that he smelled something like sour milk through the supply lines. Managers brought in an air-testing group called the watermelon gang, so named for the melon-size instrument they used to do their job. They found nothing, and the test resumed. The intercom was acting up again, as it had on previous days. For the next four hours the Apollo crew and its vast support crew slogged through the countdown checklist. Grissom complained vigorously about the bad intercom connections. He was entitled. As one of the Mercury astronauts still serving in the space program, Gus Grissom was to be included on the list from which NASA would select the man who was to be the first human to set foot on the moon.

Approaching the moment during countdown at which the spacecraft would need to switch from the space center power lines to its own internal power supply, the intercom transmitted a call about fire in the cockpit. Twelve seconds of jumbled words and yells followed. Controllers in Florida and Houston heard one last scream, then the line went dead.

The astronauts died of smoke inhalation before NASA and North American workers could get near the capsule's hatch. Explosions and smoke drove them back twice. Quickly the capsule's exterior grew too hot to touch without firefighters' gloves. Afterward, investigators disassembled the capsule into its smallest pieces. Though it couldn't be proved, a likely spot for the first spark was on Grissom's left side, down low, where a loose bundle of wiring passed by a little door for access to some of the life-support system equipment. Opening and closing the door over the weeks could have scraped off some insulation, in which case bare metal might have struck a spark and ignited something that smouldered at first in the pure oxygen, then burst into flame, lighting everything around it. What burned? With so much oxygen many things could have flared up. Fumes from the glycol coolant, perhaps, or plastic netting, or strips of Velcro that the astronauts used freely to stow their gear around the cabin so things wouldn't go adrift under the microgravity of space. In a cockpit that was supposed to have no more than five hundred square inches of plastic fiber, the walls held ten times that amount.

Right or wrong, guilty or innocent, a lot of people had to "go away." NASA reassigned its Apollo program manager, Joe Shea, to Washington, D.C. It was a dead-end job, and he left the agency six months later. North American fired, retired, or reassigned the general manager of the Apollo program, along with company managers for instrumentation, ground support equipment, project test engineering, and test operations.

A PIN LIKE A HAND GRENADE

Oxygen haunts us still, because people still fail to look outside their tunnels and take warning. The last story about oxygen concerns small, shiny canisters that make oxygen on command. They caused the strange apparition that fisherman Walton Little saw on the afternoon of May 1, 1996. It turned out to be a rear-engined passenger jet, diving toward the shallow waters of the Everglades. Little was in a fishing boat on a canal at Everglades Holiday Park. A pilot himself, Little stood up in his boat for a better view of what was happening a mile to the east. There was no smoke or flame. Tipped on its right side, the airplane hit the water and raised a mighty plume of mud, vegetation, and shattered limestone. Little's cell-phone call was the first visual confirmation that ValuJet Flight 592 had gone down.

ValuJet was the fast-growing product of federal airline deregulation. ValuJet grew from a two-airplane fleet to fifty-two airplanes in less than three years. ValuJet had started flying DC-9s in 1993. Meanwhile it was adding a new model of airplane, the McDonnell Douglas MD-80 series. ValuJet bought three of them from McDonnell Douglas Finance Corporation in early 1996 and had the airplanes ferried to

SabreTech Corporation in Miami, one of its three heavy-maintenance contractors.

SabreTech's contract called for it to credit the customer for $2,500 each day it was late in delivering the jets for service. So SabreTech managers sent the word out: mechanics would work twelve hours a day, seven days a week, until all three airplanes rolled out the hangar door. SabreTech hired mechanics on a temporary basis to make the deadline, at one point having more than seventy mechanics working on the MD-80s.

One of the jobs involved replacing outdated parts in the jets' emergency oxygen supply system. In case the cabin decompresses when flying above fourteen thousand feet, airliners carry a small amount of supplementary oxygen, for delivery to each passenger through the familar yellow plastic masks that flop down in an emergency, dangling from their white lanyards and clear tubes. Some airliners keep the oxygen as a gas in pressurized tanks. The other method, the one the National Transportation Safety Board implicated in the ValuJet crash, uses canisters with chemicals that produce oxygen when triggered. The jet that crashed was carrying dozens of expired oxygen canisters as cargo.

The canisters were of stainless steel and each was about the size of a salami: more than two inches in diameter and eight to ten inches long, with the size depending on how many passengers they were designed to serve (see Figure 7). On the end of each canister was a retaining pin. As with a grenade, pulling the pin allows a small spring-loaded hammer to strike a percussion cap detonator. This action sets off a small explosive charge, igniting the contents of sodium chlorate. In burning, the contents of the canister release a stream of breathable oxygen, good for emergencies of up to twenty minutes long. The outside of the case reaches 500°F when producing oxygen.

In this case, the problem of tunnel vision was shared by the Federal Aviation Administration. The FAA had been getting reports of oxygen canister fires at least as far back as 1986 but did nothing to ensure that airlines stopped the dangerous practice of hauling the canisters around as cargo for their own convenience. The first report of

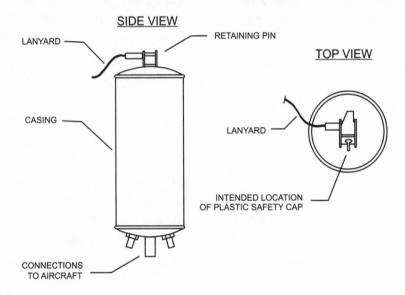

FIGURE 7: AIRLINER OXYGEN CANISTER

SIDE VIEW

LANYARD

RETAINING PIN

TOP VIEW

CASING

LANYARD

INTENDED LOCATION
OF PLASTIC SAFETY CAP

CONNECTIONS
TO AIRCRAFT

PROBABLE SEQUENCE

1. Mechanics remove outdated chemical-type oxygen canisters from two airliners without installing safety caps on old canisters removed.

2. Outdated canisters go aboard different airliner as cargo.

3. In flight, retaining pins on one or more canisters come loose.

4. Lack of safety caps allows canisters to trigger, generating intense, oxygen-enriched fire in cargo compartment.

5. Airliner crashes after fire burns through to flight systems.

Adapted from NTSB

trouble came after an American Trans Air DC-10 caught fire on August 10, 1986, and burned down to a total loss while parked at a gate at Chicago O'Hare International Airport. That fire happened because a mechanic had inadvertently triggered an oxygen canister while rummaging for spare parts down in the cargo hold.

And in October 1994 a delivery van driver inside an Emery Worldwide building in Los Angeles noticed that a box was smoking inside his van. He drove the van outside and was able to yank the box out before the fire took hold. Afterward the FAA's Office of Civil Aviation Security reported back to headquarters that someone had tried to ship an unmarked box with thirty-seven airline-type oxygen canisters to an airline in New York. The firing pin on one of them had triggered, igniting the canister and burning a hole in the box. The canisters had had no safety caps on them, and they had been packed in the box with plastic bubble wrap, with their lanyards taped to the side.

The checklist of tasks controlling the canister changeout at SabreTech was listed on a ValuJet "routine work card," and it was numbered 0069. A ValuJet employee had put the card together by reading through the manufacturer's maintenance manual. The work card said nothing about what to do with the expired canisters, though the maintenance manual said plainly that an old canister could be made safe only by firing it off.

The work card did say that a mechanic should install a plastic safety cap in the process of removing each canister from its holder. Made of rubbery, yellow plastic, the safety cap acted as a cushion over the detonator that prevented the hammer from igniting the contents at the wrong time. But the mechanics had no spare safety caps to use when they took the old canisters out, so they ignored this step and just carried the canisters out of the airplanes and piled them on a storage rack nearby for someone else to worry about. After installing the new canisters they could have gathered up the safety caps pulled from those and stuck them onto the old canisters, but no one remembered the unfinished job, or thought it important. Better yet, they could have hauled the canisters out to the parking lot and set them all off in an hour or so, but no one did that, either.

By the first week of May, five boxes of old oxygen canisters clut-
tered up the parts rack at SabreTech. The mechanics moved them over
to the ValuJet holding area, where a stock clerk, needing to straighten
the place up, got the idea of shipping them back to ValuJet in Atlanta.
The stock clerk apportioned the canisters among the five boxes as
neatly as he could, draped them in a thick layer of bubble wrap, then
sealed the boxes. The boxes had no hazardous-materials warning labels
on them, and the shipping label listed the canisters as empty.

On May 11, a SabreTech driver arrived to load up the boxes,
which weighed about fifty pounds each, along with three DC-9 tires.
He unloaded everything onto a baggage cart at the ValuJet ramp at the
Miami airport, next to Flight 592, a DC-9 twin-engined airliner. On
the flight deck were Captain Candalyn Kubeck, thirty-five, and First
Officer Richard Hazen, fifty-two. Kubeck would be handling the take-
off and climbout. She had logged more than two thousand hours in the
DC-9 and almost seven thousand hours in other aircraft. In September
1995 she had shown good sense when an air-conditioning unit had
overheated in her DC-9 and sent the smell of smoke into the cabin.
Rather than wasting time trying to figure out the problem in the air,
Kubeck had returned the jet to Dallas-Fort Worth immediately.

Hazen had sixty-five hundred hours of pilot experience and had
served in the air force. As copilot his job was to check out the airplane
and make sure that cargo and passengers stayed within the weight and
balance limits. He discussed the loading of the five boxes and tires with
the ramp agent and they agreed to store them in the forwardmost
compartment, bin 1. A ramp agent loaded them as follows: one large
airplane tire lying down, then a smaller tire on top of that, also laid
down horizontally. He piled the five boxes of canisters around the big
tire, hearing a clinking sound as he set one of them down.

At some point before the DC-9 left the ground at 2:03 P.M., the
NTSB hypothesized later, some of the canisters in bin 1 had shaken off
their retaining pins. The hammers fell, oxygen flowed freely, and the
chemical reaction heated the canister walls, which then ignited the plastic
bubble-wrap. It was in many ways a replay of the burning box of canisters
at the Emery Worldwide office in 1994, but this time it happened in the

air. Since there was no fire or smoke alarm in the bin, the first sign of trouble didn't come until it was too late. According to the cockpit recording, Kubeck heard a thump and asked Hazen, "What was that?"

Likely, it was a tire exploding. By this time the fire had been burning for at least seven minutes and had eaten through the liner of the cargo compartment, reaching the electrical wiring. Hazen and Kubeck discussed whether an electrical bus was tripping out, a problem that had happened before. It got rapidly worse. "We're losing everything. We need, we need to go back to Miami," Kubeck told Hazen at 2:10 P.M. They could now hear cries of "Fire! Fire!" from the cabin. Kubeck pointed the airplane down for a return to the airport as Hazen radioed the tower for clearance.

If Flight 592's problems had stayed only among the wiring, Kubeck would have been able to bring the airplane closer to the airport and perhaps could have saved some or all of the passengers. The DC-9's control surfaces are hydraulically powered and so the airplane can stay in the air even with electrical problems. But when Kubeck pulled back on engine power for descent back to Miami, the left engine did not respond to the throttle change. It stayed at climbout power. If left untended the airplane would quickly roll over to the right under the off-center thrust, so Kubeck countered by applying the ailerons on the wing, hard to the left. The airplane began a steep turn to the left, dropping from ninety-five hundred feet. Somehow she got the jet leveled and straightened at nine hundred feet and pointed it back to Miami International Airport.

But immediately after this, at 2:13 P.M. and 34 seconds, Captain Kubeck lost control. The jet rolled to the right, turned its nose steeply down, and dove into the swamp. The airplane and everything in it went to small pieces as it bottomed out on the limestone floor of the swamp, seven feet down. On board were 110 people, all killed instantly. Search and rescue crews never found a trace of Kubeck. In a federal criminal trial, a jury found SabreTech guilty of transporting hazardous materials and failing to provide training to its employees about how to handle these materials. The jury did not convict any employees for falsifying documents.

SAFEKEEPERS

Safety devices are the closest most of us come to seeing the aftermath of accidents at another time and place. I won't even try to tally the safety devices a typical person could come across in a single day. A panel of mechanics and experts could write down hundreds of them. Buildings have many of them, most out of sight, like the interlocks and emergency brakes on elevators. The stairways of modern buildings have a swinging gate at the first floor landing as a signal to panicked people fleeing their building during a fire; the gate is supposed to tell them to stop going downstairs at this level because otherwise evacuees will end up in the basement and not be able to fight the crush of people still coming down.

Ever wonder why the threads on propane-tank fittings are opposite the usual righty-tighty–lefty-loosey style? It prevents people from attaching the wrong fittings to a tank and putting propane in equipment that isn't safe to take it.

Gas leaking from a stove or furnace smells like rotten eggs because a safety device makes it so. Methane, the main burnable part of natural gas, has no smell. The gas companies add a smell to gas to get your attention in case of leaks. The lesson that even rotten-egg smells deserve a place in our complex world came from a redbrick schoolhouse.

In 1937 the Great Depression was still going strong, but New London, Texas, was having none of it. The city lay in the center of the fabulous East Texas oil field. This forty-three-by-twelve-mile pay zone would provide much of the Allies' oil for the coming war, sending it northeast first by tanker and then by the Big Inch pipeline. Tapping the wealth, New London had built what some people called the "world's richest school." The building handled children for a distance of fifteen miles, many the offspring of the oil field workers. Anyone gazing out the windows could see oil derricks and pumpjacks in every direction. The school had seven wells of its own, located right on the grounds. The school was big, at 253 feet long, and arranged in the shape of an *E*.

E for "explosive." In early 1937 the school district dropped its contract for heating gas with the local utility, the United Gas Company. Instead the school's workers, with the approval of the school authorities, switched over to buying untreated residue gas from the Parade Gasoline Company. Many homes and businesses did the same thing. Buying "green gas" saved the district about three hundred dollars a month.

At 3:00 P.M. on March 18, 1937, the elementary portion of the school building had let out for the day, but some five hundred teachers and students remained in and around the New London Independent School. Sometime earlier that afternoon, a gas leak had opened up in a connection for the complicated heating system. The gas had no smell and no one noticed it. The composition of the gas-air mixture in the basement and first-floor rooms rose into the explosive range. An instant after industrial-arts teacher Lemmie Butler closed a switch to turn on a sanding machine, the ground floor rose up, the walls fell in, and the roof came down.

Just four months after the New London School disaster, Texas passed a law requiring that those selling natural gas add a "malodorant" so people would smell a leak. Today one common malodorant is methyl mercaptan, a chemical so smelly that an eyedropperful would be enough to safeguard the gas supply of a very large city for days.

The school superintendent, who along with the school board had known about and agreed to the purchase of the cheap gas, lost his son in the explosion and his job afterward. Said the final report on the disaster, which cost 298 lives: "It was the collective faults of average individuals, ignorant of or indifferent to the need of precautionary measures, where they cannot, in their lack of knowledge, visualize a danger or hazard."

MINUTEMEN ON THE HIGH PLAINS

During the Cuban Missile Crisis of October 1962, apparently there was no time to visualize, either, about something that could have gone terribly wrong for the world at Malmstrom Air Force Base. During the showdown with the USSR, defense contractors and air force personnel rushed to get the new solid-fueled Minuteman 1 missiles up and working in case the president wanted to fire them at the Soviet Union. Malmstrom, near Great Falls, Montana, was the first base in the country to receive the new missiles. Each carried a 1.3-megaton warhead.

Scott Sagan, codirector of the Stanford Center for International Security and Cooperation, went through old documents from the 341st Missile Wing and interviewed people who had worked at the base. He found that the crisis arrived before things were quite ready, so the men took a number of patriotic shortcuts with the Minuteman safety rules. The system was so new and unfinished that missiles frequently came on and off the firing line during the weeks of the crisis, as urgent work was needed, then completed.

Sagan's findings should send icy fingers along your spine because back in these early years of Minuteman, an underground launch control center did not need authorization codes from the president, or anyone else, to start the process of launching its pack of ten missiles. After the Cuban crisis there would be a sanity check in place because other launch control centers nearby could veto one center trying to start a missile countdown. But Malmstrom had only one working launch control center during the Cuban crisis—hence no veto from other centers. According to the written histories of Malmstrom and official statements, commanders removed the firing panel so no mistake or poor judgment down in the underground launch control center could have started World War III. They locked the panel in a guarded vault an hour's drive away.

Or did they? Sagan found very compelling evidence that, in an effort to make the base ready to do its bit in case of a nuclear war, someone took the firing panel out of the vault and stashed it somewhere much closer to the launch center. Somewhere very close. Two of the men who worked in the launch complex at the time told Sagan that the setup would have allowed them to launch the missiles on their own, without any other capsules to issue a veto. One man said he could have launched the missiles by himself.

A few minutes of perspective, a little time out of the tunnel, might have led the hard-charging men of Malmstrom to see that, to paraphrase *Othello*, they were working not wisely but too well. The Strategic Air Command already had close to three thousand warheads at its disposal, and Malmstrom's rush added less than ten more to the national arsenal. Just one of them, unintentionally launched at the height of such a crisis—a foreseeable event considering all the electronics work on the brand-new missiles going on at a time when safety interlocks were so loose—would have been enough to ignite a war.

We began this chapter by recounting the events of the *Thresher* disaster. You'll recall that this nuclear submarine exploded under water because incoming high-pressure water compressed the air in the boat. Oil leaks provided the fuel, and the atmosphere ignited like the fuel-air mixture in a diesel engine. The *Thresher* hit its depth limit somewhere short of two thousand feet. In the next chapter we'll examine human limits.

7: RED LINE RUNNING

HUMANS HAVE A LIMIT, TOO

Just hearing the words *control room* conjures up an image of a place that by definition is secure and calm, no matter how powerful the forces being tapped just outside. The chairs are comfortable; the operators have instruments to consult at their leisure; levers and buttons keep the machines at bay. Lights banish the shadows, and walls muffle the noise.

This appeared to be the situation shortly after midnight on April 26, 1986, when a small crowd of engineers, technicians, and foremen occupied the control room at the V. I. Lenin Chernobyl power station, Reactor 4. The reactor complex was on the edge of the Pripyat River, eighty miles north of Kiev. The crew was attempting to carry out an experiment during a short transition period as the reactor came off line and coasted to a shutdown for its annual maintenance work. The goal was to squeeze enough electrical power out of a reactor-generator set during an emergency shutdown to give time for diesel generators to come on line to pick up the power loads. But the operation wasn't going well. When operators tried to send control rods into the reactor core, they stuck. Then the operators heard a rumbling noise followed by a pounding so deep they could feel it through the floor, like something big and scary had gotten loose out there in the quarter-mile-long turbine hall. At about the time the foreman decided the reactor was running out of water, the walls blew in and the roof fell. A white dust settled over the scene, and the lights went out.

The foreman of the reactor section ran into the room, reporting that the massive blocks on top of the reactor, almost two thousand of them, each weighing 770 pounds, were jumping up and down in a kind of maniac dance. Seconds later, another man rushed in to announce that the turbine hall was on fire. He ran back out the way he came, followed by two men from the control room. The first glance was startling enough—the roof had partially collapsed and small fires were burning—but the scariest sight had to be the red-hot rubble on the floor. Broken chunks of uranium fuel lay scattered across the yellow linoleum, along with burning pieces of graphite from the reactor.

These three men were the first to see just how out of control a wrongheaded experiment could go. Shortly, many more people would feel the heat. Two dozen reactor operators and firefighters died at the Chernobyl complex, thousands more probably died downwind, and at least seventy thousand people across northern Europe were contaminated by dangerous levels of radioactivity.

The straightforward explanation of the Chernobyl disaster is that in following orders to test the reactor's ability to generate an emergency margin of electricity as it coasted toward shutdown, the operators turned off the automatic safety shutdown system, thus violating safety rules. These errors triggered design problems in the reactor. After running for a year and storing radioactive decay products, the reactor was going to be most unstable when operating at low power. The operators' dogged attempts to keep the power-generation experiment going, despite all difficulties, brought this unstable machine closer and closer to a runaway. The final crisis came when workers panicked and attempted to drive in all the control rods to shut the system down. No one understood that the control rods would pose a high hazard in this situation, because the outer graphite tips of the rods were less effective at controlling the reaction than the water they pushed out of the reactor channels when the rods entered. The design error sent the reactor's output toward a new and lethal peak. Indeed the control-rod tips boosted the reaction so far and so fast that the reactor began warping and breaking up. The rest of the control rods' length, holding the reaction-stopping metals that would have contained the crisis, never got far enough into the core do the job. It's likely that the reactor suffered two separate explosions: steam explosion and an explosion that had the properties of a very small atomic bomb.

One takeaway from Chernobyl for the rest of us is that it doesn't take much of a nudge to bring about a drastic change in the behavior of a complex machine when that machine is already under great stress. The graphite tips at the end of the control rods hadn't been a big deal before because the reactor had never been so close to a runaway. But this time the circumstances turned a device that everybody expected would contain the problem, the control rods, into something that made it vastly worse.

Called the "RBMK" model, this power station reactor was derived from an old Russian military design used for making plutonium. In Russian, RBMK stands for "channel reactor of high output." High output indeed: after heat output shot to one hundred times its permitted maximum, the fuel core exploded, blew the eleven-hundred-ton lid off, and threw uranium and radioactive byproducts into the atmosphere. The number 4 reactor released at least fifty million curies of radioactivity—or maybe much more. The reactor probably dumped two hundred times as much radioactivity into the environment as the two U.S. bombs detonated over Japan in 1945.

Certainly the operators of the number 4 reactor did their part to push the machine past the margin of safety. But the whole system at V. I. Lenin, the way utility managers designed the test and the way designers set up the reactor, pushed the operators themselves across a redline limit of human capability.

You've probably seen a variety of gauges with bright red lines on their upper or far-right-hand margins. For example, tachometers, which gauge engine revolutions on cars and trucks, each have a red line. When the tach needle stays over the red line, beyond 6,000 rpm or thereabouts, it denotes that something bad is going to happen to your vehicle. It could be something as benign as the breaking of a timing belt or as major as a connecting rod being punched out the side of the engine block. For old-time steam boilers, two vital redline limits were steam pressure and water level; if the former got too high or the latter got too low, a powerful explosion was imminent. How powerful? A boiler that bursts with 125 gallons of water at 300°F releases the explosive energy of four pounds of nitroglycerine.

Today's bigger and more complex machines have many more red-line limits than do simple machines. These limits involve such matters as oil temperature, airspeed, turbine vibration, voltage, and many others. The importance of watching the redline limits, then, had better be clear to anyone wanting to survive on the machine frontier.

SHORT ON SLEEP

We humans have our redline limits, too, at which our abilities drop off dramatically. One such hazard to those working on the machine frontier is fatigue, and it's on the rise for many reasons, such as the preference for companies to pay the workers they already have more overtime instead of taking on additional staff. Companies want managers to stay as long as it takes to get the job done. According to Peter Smith of the University of Arizona's Lunar and Planetary Laboratory, one example of this practice is NASA's faster-better-cheaper approach, which places intense demands on key people. Under such circumstances, an eighty-hour work week is not unusual.

There's something to be said for showing up at weird hours. Managers who want to see how their twenty-four-hour factories are really running know they will see and hear things at night that they wouldn't during the daylight shift. As a reporter I most prefer to visit a factory, police scene, or laboratory in the middle of the night, the time that Ray Bradbury says is the low ebb of the soul and therefore the most revealing. While writing an article about a drilling rig in Kansas I made it a point to get to know the four-man crew that worked the "tour" ending at 10:30 P.M. For days I watched the driller and roughnecks at work on the derrick floor, following them into the doghouse to hear their unvarnished stories as they ate peanuts and waited for the rotary bit to gain enough ground that they could add another length of drill pipe. They even offered to teach me how to wield tongs and chain, the massive tools of their trade. Each night after work they piled into an old Delta 88 for the 140-mile drive back home to Great Bend. Arriving well after midnight, they left home at noon to start over again.

Such long hours exact a toll. It's tough on truckers who routinely drive seventy hours per week, and the automobile drivers who have to share the highway with such dangerously tired people. Seventy hours is the federal limit that truck drivers can sit behind the wheel each week, but some companies try to swerve around the rules by paying their drivers in cash or by putting a "ghost" driving partner on the logbooks. This splits the hours and brings the real driver below the maximum. One insurance-industry survey in the mid-1990s found that a quarter of truckers questioned admitted they had fallen asleep at the wheel at least once during the preceding month. A worldwide study of oil-well blowouts from 1970 through 1977 showed that half of them occurred in the eight hours after midnight. Errors spiked between the hours of 2:00 and 3:00 A.M.

According to sleep researcher Dr. William Dement, some people can stay awake for long periods if their work happens to be highly motivating. Once the stimulus goes away, though, the urge to sleep is almost uncontrollable.

Long hours are an obvious hazard; not so obvious is the safety hazard posed by shift schedules at twenty-four-hour-per-day operations. Until problems became obvious around 1980, one common factory work pattern called the "southern swing" shift had workers doing eight hours of early-morning shift for a week, then the night shift the week after that, then the daylight shift before starting the cycle all over. At least fifteen million workers in this country have to work nights or some pattern of rotating shifts. Some of the worst industrial disasters of the twentieth century started in the wee hours: Bhopal, Three Mile Island Unit 2, and Chernobyl Reactor 4, to name a few.

Good communication between crew members is one casualty of fatigue. Once I stayed up all night to hang with some technicians trying to solve a difficulty on a ship in the Gulf of Mexico; as the problem dragged on and we approached a second dawn after twenty-four hours without sleep, even in my fuzzy-minded state I could see how far the quality of our conversation had fallen, which was mostly silence broken by occasional free association. When I went for a walk around the ship to stay awake, I kept seeing human faces in the outline of the machinery, a minor but unsettling hallucination. The men did their routine jobs well enough, but no one had much energy for tackling the difficult tasks, which needed protracted troubleshooting. Fortunately our endurance run ended soon afterward when reinforcement workers arrived via helicopter. But some workers don't have that luxury. Recalling the brutally long days and nights of the Cuban Missile Crisis, presidential adviser Ted Sorensen said one of the most worrisome revelations was how sleeplessness eroded the powers of judgment.

THE CAN-DO MAN

Early-morning fatigue played a part in a very close call with a British airliner in June 1990. The central actor was a hardworking maintenance manager for British Airways; we don't know his name, but we'll call him Jones. He worked the graveyard shift at the company hangar in Birmingham, England. Since most maintenance work on airliners is done at night to avoid expensive downtime, this shift always ended up with plenty of work. Still, the workers prided themselves on getting everything done before the morning rush. On this particular night, a twin-engined passenger jet known as the BAC1-11 rolled in for work that needed to be finished by 6:30 A.M. so the aircraft could get a wash before starting the day's flight to Malaga, Spain. At thirteen years old and with 32,724 hours of flight time, the airplane registered as G-BJRT was well into middle age. Jets can last much longer; one Boeing 747 finally departed the Sabena fleet with 94,794 hours on the meter.

Jones saw from the worksheet that G-BJRT needed its left front windscreen replaced. His small crew was already busy, so although he ranked as a supervisor and didn't have to do this sort of thing he decided to take on the awkward job of replacing the sixty-pound slab of layered glass and plastic himself. It's not like replacing a car windshield, because the installation has to be strong enough to resist tons of force from cabin pressure when the airplane is at high altitudes.

Jones began the job at 3:00 A.M. He had replaced aircraft windscreens six times before, but he nonetheless read through that part of the maintenance manual quickly. Then he gathered his tools, positioned a scaffold, and climbed up to unscrew ninety bolts from the rim of the windscreen. A supervisor from the electronics section helped him pull the old glass and fairing strips out.

So far, so good. Jones had a new windscreen ready and he had the pile of original bolts, eighty-four of which he knew to be of size 7D and six of them a little longer to hold the fairings down. But some of the bolt heads had globs of dried paint on them and others had been scarred from the removal process, so Jones refused to take the easy course of using as many of the old bolts as possible, replacing only the damaged ones. He wanted to replace them all. His troubles, and subsequently the pilot's, began about there.

Jones began his search for 7D replacement bolts by bringing along one of the old bolts to the nearby storeroom. None of the bolts were marked, but Jones knew the part number he wanted and the bins were all labeled. The 7D bolt bin had only a few on hand, far less than he needed. Jones told the man in charge of the storeroom about the shortage; the man replied a little stiffly that it didn't matter because Jones should be using 8D bolts for a windscreen replacement, anyway. Jones disregarded the comment, figuring that since 7D bolts had worked before, they would work again. And even if Jones had conceded the point, the storeroom was short of 8D bolts, too. As it happens the storeroom man, and the instruction manual that Jones had skimmed over, were correct: the proper bolt to use was the 8D.

Jones left the storeroom with matters in this unhappy state. Now his airplane had a big hole in the nose, making the 6:30 A.M. deadline problematic unless he could find enough bolts, and quickly. Jones knew of a standby parts depot located two miles away, so he drove there. He was entirely on his own at this point. Most of the bins weren't marked, none were supervised, and the light in this part of the building was bad. He kept digging until he found a bin of bolts that, when he held one up in the gloom alongside his old 7D, looked to be the same. In fact the fallible human eye was at work, and the ones he found were neither 7Ds nor the proper 8Ds. The eighty-four unmarked bolts he loaded up and took back with him to the hangar were size 8C. They were one-fortieth of an inch narrower in diameter than the 7Ds he wanted.

Working single-handed off the awkward scaffold, Jones put the new windscreen in and tightened down the bolts with a torque wrench. The position was so awkward that he didn't notice that the bolts were not offering the resistance they should have as they went into the fuselage. When he was finished, of all ninety bolts, only six long bolts of the correct size actually held the windscreen fast.

The jet met its washing deadline and took off for Malaga, Spain, at 7:20 A.M. with eighty-seven passengers and crew members. As the aircraft climbed out the captain and copilot unbuckled their chest straps but left their lap belts loosely fastened. Then, with the plane passing through an altitude of seventeen thousand feet, the eighty-four undersize bolts holding the windscreen pulled loose, taking the six good ones with them. The glass flew up and over the nose. The windscreen snapped off a radio antenna in passing over the fuselage and then fell free, landing near the village of Cholsey, Oxfordshire.

A gale of air, sweeping forward from the cabin, broke off the flight deck door and pinned it against the radio and navigation console. A force later calculated at fifty-five hundred pounds pulled the captain out of his seatbelt and launched him headfirst into the window frame, thrusting him forward like the wooden figurehead on the bow of an old sailing ship. The pose didn't last long. The airstream pinned his back against the portion of the nose above the window opening, and began dragging him out the window. It was cold out there for a man without a coat, with the temperature about zero degrees and the windspeed peaking at 396 miles per hour. The captain's legs, one caught on the seat cushion and the other on the control column, held him just long enough for a flight attendant to jump forward. The attendant grabbed the captain but could not drag him back inside, folded back as he was against the upper nose.

The copilot lowered the airplane's nose to reduce altitude and cut back on engine power. The flight attendants spelled each other at the job of hanging onto the captain, though they believed the man was probably dead by now. As the captain slipped farther out they gave up on his waist but made a last stand at his ankles. The plane made an emergency landing at Southampton, eighteen minutes after the blowout. The captain was alive, and rescue workers pulled him free. He arrived at the hospital with frostbite and a broken elbow, wrist, and thumb.

"I was certainly distracted by having so many balls in the air," Jones told investigators afterward. "When you are looking at the time and wondering how the other jobs on the shift are going, your mind can wander away from the job at hand."

Besides wakefulness, humans have the obvious limits of physical endurance and musclepower. Although there are documented cases of people surviving 150°F-plus temperatures inside ships' engine rooms, as the temperatures rise a person is liable to collapse into a coma, particularly if the air is humid. When a cabin of an airplane loses pressure at thirty thousand feet, pilots have less than half a minute to get their oxygen masks on before their thinking turns too fuzzy to save themselves. As a navy pilot in World War II, my father knew to cut away a fingertip on his flying gloves; if his fingernail turned blue it would be his last chance to notice the problem and descend to a safer altitude.

Sudden incapacitation of pilots from loss of cabin pressure caused a Learjet carrying golfer Payne Stewart to crash in a field near Aberdeen, South Dakota, in October 1999. The airplane had flown on its own, pilotless, all the way from northern Florida. Military aircraft even flew alongside to investigate, coming close enough for the pilots to see the frosted windows of the cabin, but they could do nothing to help.

POISONED JUDGMENT

Carbon dioxide poisoning doesn't have as rapid an onset as oxygen starvation, but, as we will see, the effects on awareness are as serious when the people affected have to operate and fix machinery in a crisis.

H.M.S. *Thetis* was a coastal-patrol submarine of the Royal Navy, launched, tested, and sunk in the same year, 1939. Only 4 men of 103 on board got out alive. On June 1, 1939, the *Thetis* set out from Birkenhead, England, for her first diving trials. She had her full fifty-three-man crew plus fifty extra people from the shipyard and a catering firm. At 1:30 P.M. the *Thetis* stopped engines in preparation for the trials, fifteen miles north of the nearest land, the Welsh north coast. The water was 140 feet deep. The *Grebecock* came alongside to take off passengers, of which at least two dozen were not needed for the trials and therefore should not have stayed aboard. But not one of them wanted to leave, not even the two caterers. Each extra person in an emergency would consume precious air if the submarine got in trouble, but Lieutenant Commander Guy Bolus generously let them stay. It was a serious mistake. It would leave a sunken *Thetis* with only half the forty-eight-hour reserve of breathable air it normally would have had.

At 2:00 P.M., with the weather clear and calm, Bolus gave the orders to dive, with the boat moving half ahead on battery power at five knots. But filling the ballast tanks didn't send the *Thetis* down, even with the weight of fifty extra people. The boat's executive officer gave orders to fill the auxiliary trim tanks, normally used for balancing the boat rather than diving. The *Thetis* settled down enough so that the decks were awash, but then would go no farther. Bolus had the crew turn the hydroplanes to maximum diving angle. Most of the conning tower dipped below the surface, but stopped her descent there as she cruised forward at half speed.

After forty-five minutes of cruising along the Irish Sea, Bolus sent men to check whether all the tanks that were supposed to be full of water actually were full. In checking the bow torpedo tubes, torpedo officer Lieutenant Frederick Woods didn't notice that the test cocks (for checking the presence of water) were choked with paint, and so he concluded the torpedo tubes were empty. Then Woods decided to open the torpedo tubes, using the rear loading doors, to see for himself whether they were empty. He had Leading Seaman Hambrook open the tubes' rear doors in numerical order. When Hambrook moved the lever on tube number 5, the rear door crashed open and admitted a solid cylinder of cold seawater.

Later Woods said that had he been able to think things through clearly, he would have known to grab the lever that operated the number 5 bow cap door and turn it to "shut," because that was the most obvious reason the tube was flooding the submarine. (Though the subsequent investigation never settled the issue of how the bow cap door came to open, it's plausible that it had been opened back at the shipyard and was left so until the disaster.) But the need to seal off the torpedo room filled his mind instead. Water was coming in fast, driven by the forward movement of the submarine plus a twenty-foot depth of water pressure, but the men would have enough time to seal off the torpedo room if they acted instantly. The *Thetis* could, by emptying all ballast tanks, drag itself back to the surface even with a flooded torpedo room.

Woods gave the order to evacuate. There were four doors through the bulkhead, and three of them were shut already; if the sailors got the last door closed the boat would be safe. One piece of equipment needed to work well at this moment, the bulkhead door on the upper port side. But it didn't because of a cost-saving measure. Instead of the quick-closing design, which swings freely and then locks shut with an energetic, quick spin of a single wheel at the center of the door, this particular torpedo room door needed a person to fit and tighten down eighteen turnbuckles to lock it shut.

The men standing on the slanting deck in the stowage compartment pulled the door upward to bring it to the frame in the bulkhead, since by this time the boat was heading downward. The door refused to close all the way, making only a tinny clinking sound instead of the solid thump they wanted to hear. The men opened it again to see what the problem was: one of the turnbuckles had fallen loose from its temporary fastening, down into the space between door and frame. They put the turnbuckle back but the door wouldn't quite shut. As the lights

went out the control room sent along the order to evacuate the torpedo stowage room and fall back to the third compartment. Propulsion
batteries lay under the deck of the third compartment, and if seawater
got into the batteries the men would die quickly from chlorine gas
poisoning.

It must have been difficult for Lieutenant Woods to abandon the
bulkhead door, knowing that he was responsible for checking the tubes,
knowing that a few more seconds might be enough to secure the door,
and, finally, knowing that the *Thetis* did not have the buoyancy to get to
the surface with two compartments flooded. The men pulled themselves
up the slanting deck, there fighting a wave of boxes, stools, and tables tumbling down with the steep angle. Just as they got the door closed, the nose
of the submarine rammed into the seafloor at a 160-foot depth, throwing
everyone off their feet or against the nearest immovable object and leaving the submarine at a painfully steep angle. During late afternoon the
submarine settled to the seafloor. Rather than attempt escape from this
depth, the crew of the *Thetis* tried and failed three times to get into the
flooded torpedo room through an escape chamber, close the rear door to
the number 5 tube, and open two pipes that would allow a pump to draw
out the water. Then the *Thetis* would regain her buoyancy. The boat's
executive officer tried to do it alone but had to stop when he found the
pain intolerable as pressure in the escape chamber increased to match that
of the sea depth.

Fearing that men would drown before rescue arrived if he sent
them up in the dark, Bolus decided to wait until morning to send anybody out the escape hatch in the stern of the boat. Bolus was right
about the absence of rescuers above; no one was up above to pull them
out, because the tug that had accompanied the *Thetis* to the diving test
had drifted miles away under the push of wind and sea.

These efforts had used up about eight hours, or one-third, of the
total time the men could live with the finite air supply in the submarine. But because each hour that passed brought more carbon-dioxide
poisoning, more lassitude, and more chill, the men would not be able
to accomplish as much as time passed. From this point on, errors would
grow more frequent as the carbon dioxide level rose. That night the

men talked of the sinking of the American submarine *Squalus*, a little more than a week earlier. One difference was now obvious between the two submarine services. The Americans had been trained in a one-hundred-foot-deep tank on how to tolerate depth in escape attempts. The British submarine service only required men to try escaping from fifteen feet. Thirty-three men survived the sinking of the *Squalus*.

Nine hours after the flooding, the air was unpleasant but tolerable. But in the hours after midnight, talking and working became difficult, even as the shipyard men struggled to hammer out a piping system that would pump overboard fifty tons of diesel fuel and so lighten the stern, bringing it back to the surface for easier escapes. The next morning, after seventeen hours underwater, Guy Bolus allowed a new set of escape attempts to begin. Four men made it out the rear escape chamber in two attempts, but after that the air grew so poisonous that no one could think straight enough to fix a minor problem that arose with the escape chamber's outer hatch. It kept them from opening the hatch more than a few inches. The carbon dioxide levels were bad enough—causing severe dopiness, headaches, and a sense of hopelessness—and then another mishap made the air even worse. In classic one-bad-thing-begets-another fashion, some sailors had opened the escape-trunk door a little too soon on one of the escape attempts and seawater had run down the steeply tilting deck into the motor room. The salt water shorted electrical gear there and started a fire.

At last, just about at the 3:00 P.M. deadline, another befuddled mistake at the escape chamber let the ocean pour into the *Thetis* through an open escape-trunk door. The *Thetis* sunk to the bottom for the last time, foiling an attempt from the rescue ships above to hold up the stern with steel cables long enough to let a cutting-torch crew break in.

In addition to the well-known red lines that people reach under such adverse conditions as poisoned air, there are limits of understanding and perception. These differ from person to person, and special training or personality characteristics can push a person's limits, but all of us have our red lines nonetheless. These limits are part and parcel of our humanity, handed down from millions of years of eat-or-be-eaten existence in the forests and savanna. Too often operators and crews take the blame after a major failure, when in fact the most serious errors took place long before and were the fault of designers or managers whose system would need superhuman performance from mere mortals when things went wrong.

An efficiency expert named Frederick Taylor was one of the first to recognize that work needs to fit people, rather than the other way around. His time and motion studies of bricklayers in 1898 showed that a new scaffolding design—quickly adjustable in height and having a shelf to hold bricks and mortar—enabled masons to put up three times as many bricks in a day with less exertion.

What matters to us is how human limits play out in the cockpit. Here reside the operators whom we expect to run machines that get bigger, more powerful, and more complicated by the year. I use the word *cockpit* in the broadest sense, taking in airplane flight decks, ships' bridges, factory control rooms, or just about any place from which people are expected to master a machine.

Cockpit is an old and hard-edged word. It once referred strictly to a spot of dirt where trained roosters fought for gamblers' gain. Later, during the era of sailing ships, *cockpit* also came to mean a particular compartment on the orlop deck of a warship, toward the bow and located below the waterline. A ship's cockpit provided a sleeping chamber for the surgeon and did double duty as a surgical chamber following a naval engagement. So it's a little odd that *cockpit* later came to mean the pilot's nook in the original biplanes, carrying its etymological echoes of blood and battle. In technospeak the preferred term is *flight deck*, but *cockpit* lives on.

The cockpit is where the action is, so stress goes with the job. When revved up by high levels of stress, the brain tosses out everything seen as unimportant so it can focus attention on the problem. I noticed a bit of this when riding with a policeman in Austin, Texas, and a fire captain in Los Angeles, both times at high speeds through traffic. They had been talkative enough before the ensuing crises but clammed up at the outset of these episodes, and in fact no one present at these times felt like chatting.

Our tendency to move to extremes of concentration during times of emergency sometimes goes by the term *cognitive lock*. One side effect is that those on the scene of an industrial crisis may seize on an early explanation for the problem and hold to it despite all later-arriving evidence. Because their minds are made up and they want only to solve the problem, those in charge treat contradictory information as a time-wasting distraction. That was the situation during the first day of the Three Mile Island crisis, when the operators in the Unit 2 control room convinced themselves the pressurizer's water level was too high when it was actually far too low.

FEAR AND FURY

After intense concentration, the next step of stress is anger: anger at the equipment that refuses to obey and at the people who put them in this fix but aren't here to help. For a short time all of us are capable of raw fury at an inanimate object and the people who made it or put it there, if we are driven far enough. A supervisor or manager who sees the stress peaking should intervene to cool things down, say by putting someone else at the controls. In 1975, during a family business undertaking, I was given the job of driving a truck loaded with mining equipment up a forested mountainside in southern New Mexico. After I had bashed in a door against a tree, gotten stuck twice, and was making the tires smoke, my brother suggested that he drive for a little while. My first reaction was to take a death grip on the steering wheel and gearshift, but then I realized he was right. It was time to take a break; I was well past my red line and therefore no longer a safe driver.

Had Captain Edward Smith of the *Titanic* dropped in on his ship's radio room on the night of April 14, 1912, he would have seen the growing frustration of radiotelegraph operator John Phillips. During the day leading up to the collision, Phillips had barely been able to keep up with stacks of business and social messages passing between passengers and their onshore associates. Then the radio set stopped working for a few hours during the afternoon, making his transmission backlog worse. No officer intervened to help Phillips sort things out after he got the set working again, so by early evening any further iceberg warnings began taking a low priority even as the ship drew closer to the eighty-mile-wide ice pack. Phillips jotted down an ice warning from the *Mesaba* but parked it under a paperweight on his desk. When at 11:00 P.M. the *Californian's* radio operator interrupted a transmission between the *Titanic* and a shore station to inform Phillips that the *Californian* had stopped because it was surrounded by icebergs, Phillips fired back, "Shut up, shut up, can't you see that I am busy?" On the *Californian,* probably less than ten miles away, Cyril Evans did shut up; he also turned off the wireless set and went to bed.

The precipice that anger brings people to, and how easy it is to tumble over, calls to mind the Stephen Vincent Benet story "The Devil and Daniel Webster." In the story, a hard-luck New Hampshire farmer trades his soul for riches. Life is good until the Devil comes along years later to complete the exchange. Orator and senator Dan'l Webster says he will plead the farmer's case if the Devil will agree to a trial. The Devil says yes, but only on the condition he can select the jury, which the Devil does by dragging a net through Hell to bring up the worst of America's traitors and mass murderers. As the trial winds up, Webster sees that his job as defense lawyer is impossible. The Devil's handpicked judge has overruled all his objections, and the jurymen are grinning evilly. But as he is about to begin his final argument to denounce them all, he pauses. He sees that however holy his anger, it will consume him. He steps back from the brink and starts a wholly different speech, talking instead about the brotherhood of all Americans on the frontier.

One of the most evocative accounts of what fear and anger does to judgment is Norman MacLean's *Young Men and Fire.* Late on the afternoon of August 5, 1949, a team of sixteen U.S. Forest Service smoke jumpers leaped from a C-47 transport, parachuting into a rugged area of western Montana called Mann Gulch. Headquarters did not consider the Mann Gulch forest fire under way to be particularly serious, because it was contained within a hundred acres, but thirteen of the men had only hours to live. After gathering their gear and eating dinner the men hiked toward the fire. They found that the thirty-foot-high flames had jumped across to their side of the gulch, the north side. The fire was advancing so fast it was just a minute from sweeping over their position.

So it happened that even before the men lifted a tool to fight the fire, death was coming on winged feet. The men, it turned out, had two ideas about survival that day. Crew foreman Wagner Dodge began setting a fire in the tall, dry grass immediately in front of the group and yelled for the men to drop their tools and join him by falling into the ashes. It apparently didn't make any sense, and the men were on the verge of panic, anyway. The group unraveled, and they ran for the ridgeline instead, where the foliage was thinner. It was not a senseless plan, but the flames nevertheless caught up with them. All the runners died except for two, who made it to the shelter of some boulders. Meanwhile Dodge's crazy escape plan had saved him. His plan was not to move at all, and instead to set a small preemptive fire that burned the fuel out before the main flame front arrived. Waiting as long as he could, Dodge had lain down in the ashes, and the flames had roared on by.

OVER THE EDGE

Cognitive lock and anger have brought us to the limits of how much stress people can handle and still hang on to some of their reasoning power; now imagine the needle on the gauge moving well into the red zone. Psychologists call this state of mind "hypervigilance." It lasts for a short time and lies somewhere on the far side of panic. Hands shake, hyperventilation occurs, and heart rates peak. Committing a serious error under such conditions is likely. Within the grip of hypervigilance, people will forget their training and all key facts. Instead, old habits return to one's consciousness, clear and sharp.

Even perception changes, narrowing to take in only what people perceive as the greatest threat. An inflight emergency in even the most realistic cockpit simulator cannot reproduce that altered state of mind. A person has to believe, down deep in the most primitive part of the brain, that something very bad is coming and his time is almost up. A set of experiments by Mitchell Berkun at Fort Ord, California, during the early 1960s—which would now be banned on ethical grounds—documented the crippling effects that hypervigilance can have on clear thinking. It was a revelation. We know now that whereas a modest level of fear aids our thinking powers by increasing our alertness, extreme fear shuts it down.

One Berkun test sent recruits as passengers in an airplane that went into apparent engine trouble over the Pacific. As the plane lurched and dived, an officer told the men that the crash was going to kill them and suggested that the men fill out some life-insurance forms on the way down, for the sake of their families. These forms were in fact hidden tests of memory and thinking skills under duress. Another of Berkun's diabolical tests put recruits in a situation in which they believed a string of mortar-shell explosions was coming their way, then measured the men's effectiveness at taking certain steps that supposedly would save their lives if they acted quickly enough. All the experiments showed hypervigilance drastically reduced mental skills.

Our bodies are obviously reverting to an instinctual survival technique, a fight-or-flight boost that was good for dodging mammoths but causes real problems in trying to control complex and powerful machines that are on the edge of a catastrophic runaway. In my forty-five uneventful years, I've come within touching distance of hypervigilance three times. The most memorable time was the first, at my Missouri hometown, when at age eleven I first borrowed a relative's ten-speed bicycle. My cousin had warned me several times that I'd need to use the brake levers on the handlebars to stop. But I had only ridden this awkward, high-seated thing for a few hundred feet when I needed to pull up at a stop sign at a busy main street. Three years of habits acquired on my green Schwinn Flyer took over. There was nothing in the world now except coaster brakes, and this ten-speed had none. Backpedaling, brainless, unable even to turn into a neighbor's hedge, or to set my sneakers on the ground and skid, I blundered across the highway, saved only by the spacing between cars.

So if hypervigilance and its lesser stages of stress is a hard-wired vulnerability all of us have, what's a good manager to do? Advance planning, once again. Since people have different red lines in their vulnerability to stress-caused mistakes, it's best for managers to seek out people with the born ability to keep a clear head when things are falling apart.

Not many of my generation remember the names of more than a few Apollo astronauts now, but one name that lives on is James Lovell, commander of the *Apollo 13* mission. His ability to stay cool came to the fore at least sixteen years before that, as a navy ensign attempting to fly a jet fighter back to an aircraft carrier after a nighttime training flight over the Sea of Japan. A radio transmitter on the Japanese coast happened to be transmitting at the same frequency as his carrier's homing beacon, so he headed off in the wrong direction and separated from his wingmen. Then the cockpit lighting failed when a circuit breaker blew. After using a pen-light to check his instruments he looked out and saw a very faint streak of green glowing water. It was the phosphorescent trail of his carrier across the Sea of Japan, and following it brought him back to safety. How many people would have been calm enough in such a crisis to pick up such a subtle cue?

It should be clear by now that waiting to act until machines and men are hitting their respective red lines is too late to stop a disaster. Fortunately, we have been learning how system fractures start and then grow toward that day of reckoning.

8: A CRACK IN THE SYSTEM

FAILURE STARTS SLOW, BUT IT GROWS

The first sign that something was wrong with the liquid oxygen tank on the *Apollo 13* service module came on March 16, 1970, about four weeks before the launch. March 16 was the day for a countdown demonstration test. For maximum realism, crews had filled up two oxygen tanks that would supply life-support needs and help run the fuel cells that generated power along the way. The tanks were less than three feet across. In profile the tanks were reminiscent of the bombs that characters are always throwing at each other in old comic strips: a ball with a stubby cap on top (see Figure 8).

Technicians were puzzled by the fact that liquid oxygen did not come flowing out of tank number 2 as it should have with the standard "detanking" procedure, which had always worked before. Detanking

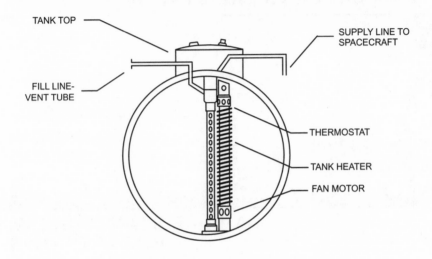

FIGURE 8: *APOLLO 13* SERVICE MODULE: OXYGEN TANK NUMBER 2

TANK TOP

SUPPLY LINE TO SPACECRAFT

FILL LINE- VENT TUBE

THERMOSTAT

TANK HEATER

FAN MOTOR

PROBABLE SEQUENCE

1. Circuitry changeover to sixty-five volts mistakenly leaves tank heater thermostat at twenty-eight volts.

2. Mishap at North American Aviation factory causes tank to strike metal shelf, jarring fill tube loose.

3. Weeks before launch, Kennedy technicians discover liquid oxygen in tank cannot be emptied normally. (Cause is loose fill tube.)

4. Kennedy technicians use tank heater to empty tank instead. Thermostat short-circuits and allows tank interior to overheat and burn off Teflon wire insulation.

5. Use of tank heater on way to moon causes spark, and combustion bursts tank.

6. Oxygen from tanks 1 and 2 leaks into space.

Adapted from NASA

was simple enough, just a matter of pumping oxygen gas into the tank through a vent tube, which forced liquid oxygen out of the tank through a second tube. Usually detanking emptied the tank of all liquid in a few minutes. But now the oxygen gas they were piping in was hissing right back out again. Something had come loose and was letting the gas take a short circuit out of the tank rather than making it push the liquid oxygen out. They had two choices. First, they could take a couple of days and replace the tank. It was a big job. If they missed the April launch window the mission would have to wait a full month for the moon to take up its desired position again.

Or second, they could improvise by turning on some gear built into the tank: electric heaters and stirring fans. That would boil the liquid oxygen (LOX) to gas and the gas would leak out through the vent in its own good time. So that's what they did, turning the heaters on and letting them run for hours until the LOX boiled away.

But in doing so the men at Kennedy missed the only chance they'd ever have to intercept a problem that came very close to snuffing out the astronauts aboard the *Apollo 13* flight. As it turned out, the only part of the mission to land on the moon as intended was the fifteen-ton Saturn-IVB third stage, which smashed into the regolith to provide information for a seismometer left on the moon by *Apollo 12*.

We know from the long history of system failures that most problems have been like *Apollo 13* in that they offered clues before the full-fledged emergency. "When has an accident occurred which has not had a precursor incident?" asks C. O. Miller, retired chief of NTSB's Aviation Safety Bureau. The answer, he says, is "basically never."

Often the warnings came far enough ahead that people had time to break the line of failure. Earlier chapters described some of those clues, like the *Challenger's* leaky solid rocket boosters. In all these, the system fractures grew without being stopped.

Aircraft manufacturers know how to intercept metal fracture by putting ribs in the fuselage to intercept cracks before they grow to a critical length. During World War II designers of Liberty ships added slots to the ships' decks, which intercepted cracks before they grew too long.

CRACKSTOPPERS

Because system fractures are sure to appear even in the best-kept settings, a good organization needs to catch and confine problems early
on. Companies need "crackstoppers" that act as barriers to a spreading
wave of failure. One example is the way that explosives plants have
learned to compartmentalize their tasks with building design and layout, so a blast in one building won't spread to any others. Thoughtful
compartmentalization can be amazingly effective in quelling disaster,
when combined with other techniques. According to the U.S. Navy,
new warships should be able to take a direct hit in an ammunition
magazine and keep fighting even so. The innovation is necessary
because the shaped-charge warheads on late-model cruise missiles are
so devilishly effective that no practical amount of ship armor can stop
them.

And the power generation industry, stung by cascading failures of
the 1965 blackout, has learned how to confine massive power-grid dis-
turbances within limited areas. It's done by putting in direct-current
high-voltage "interties" that will supply a steady source of power to a
troubled grid, but will not pass instabilities back to the rescuing grid.
All of these methods have the same idea: when waves of failure come
passing through, these barriers trap the disturbance and keep it from
adding up to a disaster.

The origin and spread of system fractures have been attracting an
academic niche of their own. NASA has come up with the term *sneak
analysis* to describe the class of problems that come and go, resisting
troubleshooting efforts because they emerge episodically out of some
subtle design error or damage during operation. Spotting important
sneaks in time can be exceedingly difficult. To take a seemingly simple
example, a late-model car with just forty switches and relays in its
electrical system can offer a trillion different permutations, like a bank
safe with a trillion combinations on the dial for a safecracker to try.
Only a very few of those circuit combinations may pose a serious
safety problem, but which ones? Altair Engineering attacked such a
problem with sneak analysis techniques, first by narrowing down the

number of switches and relays posing the greatest risk, then modeling the electrical system on a big computer for a simulation that took weeks to analyze.

All of us, myself included, have tolerated signs of sneaks at one time or another, even ones that could grow serious. At this writing I've been driving around for months with a check-engine warning light illuminated most of the time on my van's dashboard; since the vehicle still runs fine, has plenty of oil and antifreeze, and the mechanics can't find a reason for it, I've decided to let it go.

While Robert Sansone, vice president of energy engineering at Hartford Steam Boiler and Inspection Company, was visiting a fossil-fueled power plant control room for his company, an alarm horn went off. It's a very common occurrence in complex systems' control rooms. An operator reached over quickly and punched the override button to cancel the alarm. In the restored hush of the control room, Sansone asked the man what the alarm indicated. The operator asked what he was talking about. Sansone said he had just seen the man cancel an alarm by pushing a button. The operator denied it. "We went back and forth like that," Sansone told me, "until I checked the computer log and it showed that the operator had pressed the override button. Canceling alarms had become a rote." It was like the game Whack-A-Mole, wherein the player needs to bonk plastic animals as they pop out of a set of holes; the faster the reaction, the higher the score.

After one power plant failure, Sansone told me, records showed that an operator had pushed an alarm override button twenty-six times straight. As Sansone described it, "The machine finally said, in effect, okay, you win, I'll break down."

SNEAK ATTACK

The history of *Apollo 13*'s tank is a good primer on how sneaky a sneak condition can be. The problem happened in three steps, all revolving around oxygen tank number 2 and the associated hardware built into the Apollo's cylindrical service module, which lay immediately aft of the conical command module. North American Aviation was responsible for the service module and the capsule, but it bought many of the parts from other companies. Beech Aircraft of Boulder, Colorado, built the cryogenic tanks for hydrogen and oxygen to precise NASA specifications. When full, the two tanks could hold 320 pounds of liquid oxygen each, at a temperature of −297°F. Each tank was double-walled for insulation, so it didn't need any refrigeration as the liquid boiled off slowly. And it was more than just a tank. Because in outer space the tank would be cold when in shadow, designers provided special measures to ensure that enough liquid would boil off to provide a steady flow of gas: the tank had electric wires to heat, and electric fans to stir, the liquid oxygen.

The original specifications for the tanks' electrical systems called for them to run on twenty-eight volts, which is what the spacecraft's fuel cells generated. But later North American contacted Beech, directing it to raise the capacity of the tank's heater circuit from twenty-eight volts to sixty-five volts. This would be safer because the standard electrical system at Kennedy Space Center was sixty-five volts. That reduced the chance of an unintended connection to the higher voltage at Kennedy burning something up. Beech acknowledged the instruction and made the necessary changes to the heaters and related wiring—except for a single item. Someone forgot to include the tiny twenty-eight-volt heater thermostat switches in the voltage changeover, and no one along the long chain of quality control noticed the error. The thermostat was only there to shut the heater off if a hollow space in the tank reached 80°F, much warmer than the liquid oxygen. Had this been the only glitch, having the wrong thermostat, nothing bad would have happened because normally the tank heater was used little on a flight. Proof that more mishaps than this were needed to forge a chain of failure comes from the fact that all previous manned Apollo missions had gone up with the same underrated thermostat switch and they had experienced no problems.

Tank number 2 passed all its tests at Beech Aircraft and traveled to North American's plant in Downey, California, to join the rest of the service module's equipment. In June 1968 North American took two of the oxygen tanks and built them into an assembly called an "oxygen shelf" that would be part of *Apollo 10*'s service module.

Except that, and here is where things started to go seriously wrong, later NASA wanted the shelf to come out again, to be set aside for upgrade work. It would go aboard a later flight. The reason was that pumps on the tanks had been causing electrical interference on earlier Apollo missions. So on October 21, 1968, North American made ready to pull out the oxygen shelf from the service module. This shelf held two oxygen tanks and a dense web of wiring and tubes. Technicians disconnected all this from the service module. They were also supposed to remove four bolts holding the shelf to the service module, so a crane equipped with a special lifting arm could raise the loosened shelf and slip it carefully out. But someone forgot to remove one of the four bolts securing the shelf. Instead of lifting loose easily, the shelf tilted up on one edge, the lifting arm broke, and the shelf flopped back into place.

Consternation followed, but not for long. North American technicians jotted the incident onto their paperwork and looked over the tank. They saw nothing wrong. It had only fallen two inches. They would have had to take the thing apart for them to see that a "fill tube" inside the tank had broken loose; later it would cause the leak that would start the chain of events leading to *Apollo 13*'s crisis. A likely explanation is that when the shelf tilted up, the top of oxygen tank number 2 had struck the bottom of the hydrogen tank shelf just above, and this shock knocked the fill tube loose.

As we've seen, when the service module got to Kennedy this oxygen tank wouldn't "detank" on command, and the space center used the heaters to boil the gas out, not once but twice. It would have horrified them to look inside somehow and see the damage.

Imagine that your toaster switch gets stuck one morning while you are down in the basement. Probably the only damage would be some burned toast and a smoky house, because you'd notice the problem soon enough from the smell. But if your toaster got turned on without your knowing, down in a cupboard, you could be in trouble as the toaster heated its surroundings and burned the insulation off its own wires. That's one way to describe what happened to the heater inside the oxygen tank. When the temperature first hit 80°F early in the improvised detanking procedure, the thermostat switch had tried to turn off the heater but hadn't been able to do so. When the twenty-eight-volt switch tried to open and cut the circuit, the sixty-five-volt current from the Kennedy power supply welded it shut. The juice kept flowing to the heater.

Running full blast for hours, the tank heater attacked Teflon insulation on the nest of copper wires inside the central tube of the tank. A technician at Kennedy was watching over the improvised detanking setup, and he had a gauge to show the temperature inside the tank. But his thermometer only read up to 80°F because that was as high as the temperature was ever supposed to go.

Based on results from experiments afterward under identical conditions, the actual temperature near the top of the tank touched 1,000°F. It was a foreshadowing of one of the many problems that would bedevil the operators at Three Mile Island, where during the crisis a computer refused to report any temperatures higher than 280°F for water coming out of the pilot operated relief valve. There, it would also mislead operators into thinking that the reading was the real temperature.

Days later, when managers were going through the final flight review, the odd behavior of tank number 2 came up for discussion. The people considering the issue blamed the detanking problem on a recent change in procedures, because they hadn't heard about the damage the tank sustained at North American's Downey factory. They talked over the possible effects of a loose fill tube, but they showed no concern about the implications of using the heaters and fans so much. They decided that the tank would be good enough.

Apollo 13 launched on April 11, 1970. Almost fifty-six hours into the mission, a master alarm went off, signifying that hydrogen pressure was low. Houston cleared Jack Swigert to heat and stir the cryogenic tanks. Swigert had stirred the tanks twice during the flight so far, but he had not turned on the heater yet. The power flowing into the tank struck a spark, apparently setting the remaining Teflon insulation on fire in the rich oxygen environment. So it happened that, once again, plastics were burning in oxygen, three years after *Apollo 1.*

The fire burned without ado for about a minute, raising pressure and temperature, and possibly igniting aluminum in the upper part of the tank. Then the neck at the top of the tank failed, dumping oxygen into the equipment bay on the outside of the service module. That sudden pressure wave, or possibly another quick fire in the bay, blew out the bay's exterior cover. The first explosion broke open the piping shared by both oxygen tanks and dumped all the gas into space, in an unintended "detanking" procedure similar to the one that had caused the problem in the first place. By losing its oxygen, the spacecraft also lost its power supply—except for what remained in the batteries on board.

While the TV audience thrilled to the way that mission controllers and Apollo crew beat the odds and got the ship back home by rigging up brilliant solutions with the electrical system, navigation, LEM engine, and air purification canisters, the *Apollo 13* drama was not something to cheer about. This highly avoidable mishap later strengthened the cause of politicians who opposed Apollo spending and who wanted to end the program before the full suite of flights. And *Apollo 13*, or its predecessors with the same thermostat problem, could easily have been a complete disaster had the blowout occurred under slightly different circumstances. The lesson that Apollo brought back to Earth is that a quick work-around—in this case the brainstorm that running the heater could empty the oxygen tank fast enough to make the launch deadline—is a good idea only for a machine in which (a) failure doesn't matter much or (b) somebody first figures out the full effect this method is going to have on the machine.

Maybe we haven't had enough experience with spacecraft yet to develop good intuition. People allowed to work in a specialty long enough, like firefighters, can develop something very close to intuition, according to researcher Gary Klein. He calls this skill "naturalistic decision making." Such a person can quickly come up with reasonably good tactics in an emergency by fitting the facts into one of his mental models. When the fit isn't perfect, he improvises off the main path, while acknowledging to himself that he is on untried ground. Such a person knows how to pick up subtle clues that something important doesn't fit the usual models and therefore the conventional tactics might be dangerous.

Klein collected this story during his fieldwork: A fire lieutenant took an attack crew into a house to fight a kitchen fire. As they worked a hose from the living room, the fire did not damp down as it should have under the water spray, considering the fire's size. And the sound of the blaze was "hollow." Feeling uneasy, the officer ordered his men out, abandoning the house to the flames. A minute later the living room floor collapsed into a large fire that had been consuming the basement. Had they been standing in the living room at the time, all the men would have died. The lieutenant said later that he didn't know the house even had a basement, nor did he know there was any fire underneath the floor. He had just matched the facts against his experience and didn't like what he'd seen.

EARLY WARNINGS

We can hope that people will develop a similar insight when supervising complex and powerful machines. To this end, there's the matter of Brian Mehler's timely deduction when standing in the noisy chaos of Three Mile Island Unit 2's control room that the pilot operated relief valve had stuck open.

Another hopeful instance of intuition occurred during the collapse of the Baldwin Hills Dam, where a couple of vigilant workers saved hundreds and perhaps thousands of people. Baldwin Hills was a 160-foot-high dam of earth and concrete, built in 1948 in the hills above Los Angeles. It was 550 feet long and held back a lake nineteen acres in size. Late in the morning on December 14, 1963, a caretaker named Revere Wells was inspecting the dam when he heard the sound of water rushing through a drain. Wells knew that little leaks can foretell big problems with a big earthfill dam, so he climbed into an inspection tunnel to investigate. The three-hundred-million-gallon reservoir behind the dam was leaking out, and badly; investigators later said it was the result of soil settlement. Wells left immediately and called his supervisor, who investigated and confirmed the problem. This triggered an emergency plan, which consisted of notifying the police for the area downstream, opening gates to drain down the reservoir, and seeing if the leak could be stopped or at least slowed to help the evacuation. The city's chief engineer looked over the situation at 1:35 P.M. and estimated with astonishing precision that people downstream had two hours to get out. After that time the hole would be big enough to weaken a great section of the dam and unleash a great wall of water. As the police began evacuation, those at the dam used the time in a daring attempt to throw sandbags into the crack, now visible on the upstream face of the structure. Men went down on ropes to do the job. Shortly before 3:30 P.M. the engineer called to his assistant, "Get the men out— we have lost the race." The dam broke at 3:38 P.M.

Certainly we can make better use of our vast army of construction workers, technicians, and maintenance people. They're likely to see dangerous flaws developing before anyone else does, and some alertness training in this area can help them realize when to get worried. Such alertness might have helped prevent the *Alexander Keilland* platform disaster that killed 123 people in 1980. Investigators found afterward that a shipyard painter must have come face-to-face with the long crack in the steel support that eventually broke in a storm, because they found paint in the crevices where the failure had started.

People need to have wall-to-wall knowledge of a system before they start disconnecting safety systems in an emergency. The navy calls this desperate measure a "battle short." In submarines, a battle short adopted after the *Thresher* disaster allows the crew to stop an automatic reactor scram and force a submarine reactor to stay on-line if they urgently need propulsion power for survival. When used by the right people and in the right circumstance, a battle short can stop a system fracture from spreading. However, a good crackstopping approach shouldn't come to that very often.

To see how a crackstopping approach works with people, consider Cessna Corporation, which motivates its employees to catch and report problems as soon as possible. When Fred Albin started work at the Cessna light plane factory in Independence, Kansas, he joined a team putting together model 172 Skylane four-seaters. One of his tasks on the assembly line was to wield an air-driven tool called a reamer, trimming an opening in the engine cowling. One day soon after he joined Cessna, Albin ran his reamer a bit too far and gouged the fuel filter. His split-second mistake had ruined a $375 component.

In the old-time factory an assembly worker, whether veteran or tyro, would have known just what to do. If no one was watching and the damage was concealable, he would have proceeded as if nothing had happened. Whatever occurred later was somebody else's problem. Not Albin, because Cessna had trained him differently. Albin walked up to the parts desk and admitted his error. The parts man handed him another fuel filter. "If you've got a problem at Cessna you're supposed to stand up and say so," Albin told me at the plant. "It was no big deal."

Cessna also shows why a crackstopper company sets up more than one line of defense. A worker might not take up the company's invitation to admit a mistake, like leaving debris in a wing or fuselage, or she might not notice the error. In that case, somebody is going to have to discover the problem before the customer suffers the consequences. So all Cessna's assembly lines have workers assigned to descend on airplanes approaching the end of the assembly line, looking for "foreign object debris," meaning discarded scraps, lost parts, or forgotten tools. They probe and peer into the tail and under the flooring, wielding little flashlights and dental mirrors. They pound rubber mallets on wings, listening for anything bouncing around inside, and they have the authority to order a wing cut open in order to remove the debris. Hard experience says that sooner or later a loose piece at the bottom of a fuselage or inside a wing will lodge in a bad place, such as a pulley for a control cable.

FIVE HOURS AT HARTFORD

No such crackstopping measures were in place when construction workers got an inkling of what would turn out to be an extremely close call in Hartford, Connecticut. Snow flurries blew as fans gathered for a basketball game between the University of Connecticut and U Mass on the evening of January 18, 1978. The location was the sports arena at the Hartford Civic Center Coliseum. The game drew five thousand people, about half the arena's capacity. Any fans gazing idly upward, past the scoreboard, would have seen the "space frame" roof structure. This roof sat on four massive concrete supports that held it eighty-three feet over the floor, and it measured 300 feet wide by 360 feet long. It was a massive latticework of thick steel tubes in a triangular layout. It was the second largest space frame roof in the country and was five years old.

Had anyone actually climbed a ladder to measure the dimensions of this engineering wonder, acting out of some excessive curiosity perhaps, she might have found that the roof was sagging across its span—a little more so as the wet snow thickened that night. Five hours after the fans left the arena, with a little more than four inches of snow accumulation, the fourteen-hundred-ton cement and steel roof crashed down on the seats.

The problem, said the city's consultant afterward, was that the roof structure was complicated and had been improperly designed, leaving it too weak to handle the loads. The consultant said that even without the snow, the completed roof with all its weatherproofing was 750 tons heavier than planned. Most disturbing was that workers had reported that the roof was sagging at least 50 percent more than the engineers had estimated, but nothing had been done. Workers also knew that some roof panels, designed to be bolted to the frame on the outside, didn't fit the holes. They had been tack-welded instead.

Now, keeping in mind the very close call at Hartford, consider these two situations. First: The upper deck on a city-owned stadium is finished and ready to occupy but, three days before the big game and with the city under intense pressure to use the new deck, an outsider says he sees a serious safety problem with the engineering calculations. Second: an American company learns confidentially that the roof on a new factory in a developing nation might crash down on hundreds of employees. In both, the signs of impending trouble are ever so faint, more subtle than at Hartford.

Do we expect employees in a government bureaucracy and a big multinational corporation to do the wrong thing? Of course. We expect they'll cover up traces of unpleasant news and do whatever it takes to make the workday go easier. Hang the consequences. Except it didn't happen that way. One case appeared briefly in the news forty years ago. The other was never publicized, as with most near misses.

THE EMPTY DECK

Television viewers tuning into the NCAA Game of the Week football game on Saturday, October 21, 1960, saw something rather extraordinary. Bear Bryant and the Crimson Tide had come to their home field for the big opening game with rival Tennessee, setting a new attendance record, but the Alabama stadium's vast new upper deck, 8,632 seats in capacity, was nearly empty.

Announcer Keith Jackson, then in the early stage of his ABC television career and not yet nationally famous for his "Whoa, Nellie!" cry, called viewers' attention to a dozen or so people sitting in the upper deck, alone in lofty grandeur. The camera closed in on a few policemen sitting with the city's chief building inspector, then panned over the groundlings below, where fans occupied three thousand folding chairs. Other fans sat on planks laid across half the aisles and edged right up to the gridiron.

On television, the empty upper deck looked like the kind of arbitrary stunt that some ancient emperor might have pulled, clearing a third of the stands in the Circus Maximus before letting the lions loose, but the reasoning here was deadly serious.

The city building inspector who sat in the upper deck that Saturday, Myron Sasser, believes to this day that the deck was plenty strong enough for a full load of fans, with safety to spare. But what the consulting engineers saw in the plans gave them cold shivers. Contemplate a massive steel platform with eighty-six hundred people crashing down onto concrete stands holding about the same number of people.

Discovery began about two weeks before the game, when a young engineer named Dick Alexander visited the stadium. The original concrete lower decks, shaped like a horseshoe around the playing field with the end zone at the north left open, dated to 1927. Now Legion Field had a new upper deck curving over the east side and bringing the total seating capacity to fifty-two thousand. Alexander was at the stadium on behalf of his employer, O'Neal Steel, which had prefabricated the steel for the new deck.

Alexander did not like the look of the diagonal wind bracing for the new structure. These were metal struts that connected the lower concrete stands with the steel seating above to provide extra stiffness and keep the deck from swaying in the wind or with crowd motion. The struts were supposed to be pulled tight. Instead six of them were bowed out, vibrating in the breeze. Not only would bowed braces be of questionable value in stabilizing the deck, Alexander worried that they might be the sign of something much more serious going on. He mentioned the clue to his supervisor and got permission to take the issue outside the company. Alexander's brother worked at an engineering firm, so Dick called him. The informal request to check out the stands landed on the desk of Wallace McRoy, a structural engineer there.

McRoy first put in a courtesy phone call to Carl Wilmore, the design engineer on the upper deck. The two men belonged to the same church in Birmingham. Wilmore said he was fine with McRoy taking a second look at the upper-deck design, and he sent over a stack of drawings. McRoy began working through the pages, checking steel dimensions and slide-ruling his own stress calculations.

McRoy called Wilmore back, greatly concerned. The loose wind bracing was not serious in itself but he did see two problems. Without going into technicalities here, one problem pointed to the possibility that the front portions of the deck, the "cantilevered" parts that jutted out over the columns, could hinge down and break off, in the same way that an overhanging roof on the front porch of a house might snap off after a heavy snow. The other problem was greater in its magnitude—a weakness in the main support truss that ran the curving length of the upper deck. A failing truss could drop the whole deck onto the concrete stands. Thus, the problems that McRoy saw were errors in design, not in the fabrication or erection of the steel.

McRoy's conclusion, coming less than a week before the opening game, threw the city's Park and Recreation Board into a frenzy of activity. The board ran the stadium for its owner, which was the city of Birmingham. Mayor James Morgan marshaled a squad of welders and set them to work on a twenty-four-hour schedule, tacking steel plates to trusses and beams. Late on Friday night before the Saturday game, as work finished up with only hours to spare, McRoy arrived at a meeting called by the city and Wilmore's engineering firm. Wilmore asked McRoy whether he would go along with opening the upper deck now. "He put me on the spot," McRoy recalls. In the awkward pause, Wilmore's junior partner, Jim Hudson, interrupted. Hudson said it wasn't fair to ask McRoy to okay the repairs without time to check the work.

And so, despite heroic efforts to get the deck ready for the big day, the Birmingham park board decided to keep fans off the upper deck until all calculations had been double-checked by a second engineering firm. Trucks hauled over thousands of folding chairs from the auditorium. The mayor-elect worked all that night with volunteers, unloading and sawing fourteen hundred planks to improvise seats across the aisles. The planks were a safety violation, as they later found out.

In the end Wilmore, Hudson, and Luke, the original engineers, accepted "full moral, technical and financial responsibility" for the problem. One of the partners told the park board that a mathematical error had made the main truss too small to handle the load.

"I never did go out there," McRoy says about the stadium crisis now. "It was that clear in the drawings. In my opinion it would have collapsed."

Skeptics might regard the Alabama case as not proving much toward the proposition that people are going to take action for the greater good. After all, Legion Field was a stadium that friends and family of the parks board members would be using. Could it be that only a direct stake in the outcome keeps people alert to subtle signs of machine trouble and presses them to sort out the problem before a disaster?

CRISIS AT CITICORP CENTER

Fortunately for those of us who don't have relatives in command of the machines and structures we use, there's evidence to the contrary. It comes from several close calls. The most famous now is a 1978 crisis at the Citicorp Center, kept secret at the time but revealed later in a *New Yorker* article. The excitement began after the building was finished and occupied, when the structural engineering consultant for the job,

William LeMessurier, was examining the as-built plans to answer an engineering student's questions. As his look back at the work proceeded, several things surprised and then chilled him.

Citicorp Center is a steel-framed building, seemingly standing on 114-foot-tall stilts positioned at the midpoint of each side (see Figure 9). LeMessurier found out that, for one thing, the structural steel had been joined differently than his firm had planned: it had been bolted instead of welded. Bolting was an acceptable way to do things, but not as strong as welding. After redoing his calculations and ordering additional wind tunnel tests, LeMessurier realized that a combination of unanticipated factors had gravely reduced the planned strength margin of the skyscraper, most severely at the bolts connecting the steel crossbraces on the thirtieth floor.

The problem was the ability of the bolted connections to resist stress from quartering winds, which are winds that hit the building's glass-faced sides at an oblique angle. Given how the building's structure was classed, the New York building code did not require the analytical approach that LeMessurier found so revealing. LeMessurier decided there was a 50 percent chance that the whole 914-foot tower would collapse in the kind of hurricane that, on average, sweeps Manhattan Island every sixteen years, if those winds happened to strike during a power outage that would knock out an antisway device at the top of the building. Called a tuned mass damper, it required electricity to operate.

LeMessurier contemplated, ever so briefly, destroying his notes or even killing himself, since one of the contributing factors was that his company hadn't calculated the full stress on the structural connections under certain wind conditions. Instead, LeMessurier blew the whistle on the whole thing. Citicorp authorized a no-stops effort to reinforce the building's structure before the autumn hurricane season began, while setting up emergency precautions. One precaution made sure the tuned mass damper would not fail under any conditions, and another prepared to evacuate an area of Manhattan for ten blocks in all directions in case a collapse was imminent.

Teamed with carpenters who broke away office walls to expose the steel behind, welders labored through nights and weekends to add

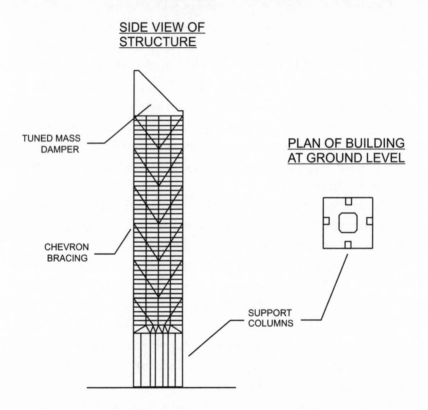

SIDE VIEW OF
STRUCTURE

TUNED MASS
DAMPER

PLAN OF BUILDING
AT GROUND LEVEL

CHEVRON
BRACING

SUPPORT
COLUMNS

SEQUENCE

1. Consulting engineer learns after building's construction that joints in steel frame are not as strong as planned: are bolted instead of welded.

2. Engineer then learns that quartering, or diagonal, winds will bring more stress on connections than originally allowed for.

3. Owner strengthens frame by welding steel plates to strengthen connections.

Adapted from building plans

enough steel to secure the strength of critical portions of the frame, particularly on the thirtieth floor. Over the weeks, workers trundled tons of massive steel plates into the loading dock and up the freight elevators, apparently without triggering any comment among the building's ritzy occupants.

A key figure in the repair was structural engineer Leslie E. Robertson, who served mainly as Citicorp's expert and who specified the exact changes to be made, based on LeMessurier's calculations.

At age thirty-four, working seven days a week, Robertson had designed the structure of the twin towers of the World Trade Center in New York; by 1979 he had his own company and had added other world-class structures to his portfolio. I found that Robertson does not look back on what the *New Yorker* called the "Fifty-nine-Story Crisis" with any nostalgia, though after the story went public the case study became famous in the field. It's a mainstay of courses that teach ethical behavior to engineering students.

"I walk a few blocks out of my way to avoid passing Citicorp Center," Robertson told me, saying that he wishes he could get the time back that he'd had to spend on the project. He regards the noblest figure in the story as Citicorp, because of the way the bank's executives agreed to spend millions of dollars fixing the building first, worrying about who would pay for it later.

Robertson is a man with zero patience for error but also one with genuine consideration for the ordinary Joe who goes into his buildings. In 1993, Robertson was called in to check the World Trade Center for structural damage from the terrorist bombing. He was one of the few civilians granted free run of the towers' parking levels during the investigation. While making his inspections, he realized that the owners of the smashed cars down there would never see them again. So he took snapshots of as many car hulks as he could and mailed them to the owners after finding their addresses through the motor vehicle department. One family received a stack of commemorative plaques that Robertson pulled from the ruins of their car. He keeps a memento of the investigation in the backyard of his house in Connecticut: a fifteen-foot-long steel beam twisted by the blast.

THE ROOF THAT WOULDN'T DRAIN

To support his point that sometimes corporations stand up and do the right thing, Robertson told me about another case he was involved in. In mid-1971 John Davis, a Latin American real estate manager for IBM's international subsidiary, made a routine stop at a new factory in Sumare, a small community in southern Brazil near the city of São Paulo. IBM had built the plant, a third of a million square feet in area, to assemble typewriters and computers.

Standing in the main assembly area at the Sumare plant one day, Davis looked up and saw several men with tape measures crawling along the steel latticework supporting the roof. The structure was a space frame, conceptually like the one to be built later at Hartford Coliseum, but not as high off the ground. Davis learned that the workers had climbed up there to take measurements of the new roof. He inquired further: a "service penthouse" structure sitting atop the roof near the center, built as a shelter for heating and ventilation equipment, wasn't working right. Water that should have drained away from the penthouse and off the roof wasn't going anywhere. According to the crew's measurements the roof had sagged two inches, defeating the slight incline built into the drainpipes. There was water sitting in the gutters, and the puddles had tadpoles in them.

Knowing that this was the type of problem that the Brazilian construction contractor should address, especially as IBM was still in the process of occupying the factory, Davis called headquarters in New York. He reached John Novomesky, the manager of area operations for IBM's international subsidiary.

Novomesky could have spent a half hour writing to the Brazilian contractor, to the effect that it had better get the penthouse pipes working right. In a couple of days, or maybe a couple of weeks, a handyman with a toolbox would have climbed up to the roof at Sumare and propped up a few pipes to get the water flowing downhill again. And, if something more serious was going on and the roof fell in, Novomesky would have had a plenty strong memo planted in the file to cover himself. Wouldn't the memo have urged corrective action in the strongest possible terms? He would have been doing a good desk-keeper's job.

Novomesky didn't do it that way, maybe because he was an engineer himself, and maybe because he'd once encountered a roof problem at a factory in Milan, Italy, that taught him when massive things hanging over people's heads are not behaving according to plan, it's better to pursue the matter more aggressively. Novomesky called a coworker and got the name of a consulting engineer, Les Robertson. The same day that Novomesky called him, Robertson got on a plane to Brazil. "It was absolutely astoundingly fast," Novomesky says now. "It was one of the most amazing things I've ever seen."

Robertson spent two days in Rio de Janeiro trying to find someone who knew something about the problem. "The architect wasn't available and nobody at the IBM office in Rio would talk to me," Robertson says. "They couldn't even say where Sumare was." But rather than giving up in disgust and heading back to his appointments in New York, he rented a car and driver and started off in the general direction of the factory. Bluffing his way into the compound without credentials, but unable to speak Portuguese, Robertson walked the hallways until he heard an American speaking Spanish. And that's when he started to get answers.

Robertson met with Novomesky for two days after he got back, laying out his findings. Some things about the space frame roof, particularly the connections secured by bolts and washers, had him worried. Robertson said that if the custom-made steel bolts were really the "high strength" composition that the builder claimed they were, the roof would probably be okay. If the bolts were made of standard steel, though, the building had a serious problem.

A representative of the contractor replied that a test of the bolts' steel would take weeks. If this was an attempt to make Novomesky go away, it didn't work. Novomesky called the vice president of finance for his company and said he wanted to clear out the Sumare plant immediately, rather than wait for results of the steel tests.

"I learned that if you want to get attention at IBM," Novomesky recalls, "just say 'I want to close a plant.'" In short order Novomesky found himself scheduled to talk with the company's chief executive, chief financial officer, chairman of the board, and top lawyer. Over the course of that meeting, Novomesky recalls, he fielded many hard questions, but the subject of how much the repair would cost, in cash and lost manufacturing time, never came up. He left the meeting with approval to clear the factory floor of occupants, pending some resolution of the problem.

After the decision to close the plant, the builder found a way to test the bolts on short notice. Robertson learned that, as he feared, the bolts were not made of high-strength steel. It meant the steel was already near the limit of its strength, in the manner it was being used at Sumare. It was significant because when one part of a space frame fails, the other pieces of the frame nearby have to take up the load. When all the connections are equally weak, even a small problem is likely to cascade, bringing down large sections of the roof. Robertson told IBM and the builder that short of replacing the roof, the best thing they could do would be to weld every one of the forty thousand space frame connections.

IBM made repairs a top priority, evacuating hundreds of workers pending action. The builder braced the roof and brought in welders from a forty-mile radius to repair the connections, fixing one section of the plant at a time. Robertson recalls flying down to Brazil regularly over the next three months, sometimes weekly, to check on the repairs until they were complete. Novomesky says he is sure that the contractor never sent a bill for the work and the whole affair concluded outside of the public eye, like many near catastrophes. There are no doubt thousands more examples of instructive close calls all along the machine frontier. It's a shame we don't hear about them—their lessons are cheap at the price. A few companies, like Premiere Manufacturing Corporation, have set up safety programs specifically based on the analysis of near-miss reports.

Near-miss reports can help us puzzle out which anomalies are tolerable and which need immediate attention. Near misses could be to safety what informants are to the police. Just as experienced detectives know they could never crack most criminal cases without the inside scoop that their informants provide, near-miss reports are often as close as we'll ever get to seeing early enough the paths of failure inside a massive system.

Of course there's such a thing as too much information, too many "open issues" all being worked at once. Which, if any, is truly serious? The difficulty of how people are supposed to detect real problems through the background noise of numerous, minor difficulties rarely gets the attention it deserves in post-disaster commentary. The company or agency involved hardly wants to advertise that it had other safety problems at the time. Although the media may hear reports about other simultaneous problems beforehand, it's just a distraction to a reporter looking for a clear and compelling theme, namely, that management had seen early signs of a fatal problem but had done nothing.

We've seen how crackstopper companies and agencies are empowering and motivating employees to report problems before they get out of control. But many worthwhile efforts, including the use of near-miss analysis, last for a year or two, flicker, and then die out. We need to go deeper into the frame of mind it takes to keep a crackstopper attitude over the long haul.

9: THE HEALTHY FEAR

ALIVE AND ALERT AT DANGER'S EDGE

How far could you run in fifteen seconds if you were very, very motivated? Ron Hornbeck got the chance to achieve his personal best on May 24, 2000. On that day Hornbeck was operator on duty at the "batch nitrator" building at the Dyno Nobel dynamite factory near Carthage, Missouri.

Hornbeck's job was to oversee a set of vats and pipes that was brewing up an oily explosive liquid called MTN, short for metriol trinitrate. The white, wood-framed nitrator building on the hillside that he worked in didn't look like much, because it wasn't the kind of place you run visitors through, and it was expendable, anyway. Just uphill was a rampart of earth and timbers, installed to shield a second nitrator building in case of explosion. The blast wall was ugly, and the workers made fun of it.

What happens inside a nitration building? *Nitration* is a chemist's term for mixing things with nitric acid. In the case of MTN, Hornbeck was taking a blend of water and two chemicals called trimethyl ethylene and diethylene glycol, and then mixing that with nitric and sulfuric acid under carefully controlled conditions.

Nitric acid is vigorous stuff that adds explosive qualities, in the form of oxygen and nitrogen "nitrate groups," to a wide range of otherwise nonexplosive substances like cotton or paper. Nitration has been called the single most hazardous chemical process in the world. Think *nitroglycerine*, a word that inspires an almost instinctual dread. MTN, though it lacks the fame of nitroglycerine, is equally liable to detonate with shock or heat.

Hornbeck knew all that, and so he kept close watch over the temperature in the nitrator. But the nitration step, the most likely source of trouble, went well. Hornbeck drained the freshly made MTN into a separation vat, where the heavier, unused acid could settle by gravity to the bottom of the tank. After a while he tapped off the acid layer, leaving about seven thousand pounds of MTN ready for the next step, a thorough washing with water. But in that little space of time, something started to go wrong in the inscrutable way of high explosives. The ideal temperature of the oily liquid in the vat was 47°F, but it was going up.

Hornbeck watched the reading climb to 55°F. On the alert now, he had several choices. He could bail out and leave the building, but things were not out of control yet. He could change the equipment setup to dump the MTN into a "drowning tank," a vat of water that (usually, but not always) will stop a runaway reaction by cooling the explosive quickly. Or he could open a big outlet that would send the liquid flowing down an open trough. Using the trough might disperse the heat well enough, and most importantly, it was fast because the equipment was already set up that way. In this situation, fast is good.

With the vat temperature now at 70°F and still rising, Hornbeck opened the gate wide and the oily MTN began to run out. He watched the tank thermometer for signs that the rate of heating was accelerating. This would indicate a runaway reaction. Heat builds up, which causes more chemical reactions, which turn out more heat. In a runaway reaction, things proceed faster and faster, as with a boulder rolling down a mountainside.

As the temperature rose past 100°F and reddish-brown fumes

gathered, Hornbeck realized it was too late in the day to try anything else. He ran through the doorway—doors swing open easily in the nitroglycerine business—and sprinted uphill, knowing that the building was designed to erupt downhill.

Knowing when to stay and when to go is a hard lesson from the past. In March 1884 Lammot du Pont was in a company building near Gibbstown, New Jersey, when a batch of nitroglycerine started fuming with red smoke. Du Pont, who had personally headed up the addition of dynamite to the family's thriving gunpowder business, succeeded in dumping the nitroglycerine into the water tank. But the liquid blew up anyway, killing him and five workers.

As Hornbeck sprinted from the batch nitrator, flames shot out of the building. The building exploded when he was less than two hundred feet away, leaving him stunned but unhurt. The blast threw debris into the pasture across Center Creek. It broke windows and shredded the siding off nearby buildings. In collapsing, the earth and timber barricade saved the other nitrator, and no other buildings full of explosives went off in a chain reaction.

It may be surprising that more than three tons of a very powerful explosive went off in the middle of a dynamite plant without triggering any other chain-reaction blasts and without killing anyone. It's because longtime explosive makers like Dyno Nobel have learned to live with fear and face their demons squarely.

FACE THE FEAR

Over the decades, manufacturers have studied the causes of plant explosions closely and shared the lessons with each other. They learned, for the most part, how to prevent them and how to keep a small blast from propagating into something titanic. Certainly they try to cut risks wherever possible, but they also know that the only way to guarantee safety would be to stop the work entirely. Their unblinking view of that last hard core of danger is a working style that some other enterprises outside of the explosives industry would do well to adopt. Some already have, like companies that use helicopter-based crews to maintain live high-voltage transmission lines. It's comparable to the more well known example of navy crews who have learned how to land heavy aircraft on pitching carrier decks, under all weather conditions, at one-minute intervals. Or, farther back in time, it's comparable to experiments by British aviators before World War II on certain air defenses, which required Royal Aircraft Establishment scientists and pilots to fly their airplanes directly into steel cables hung from barrage balloons. These rank among the most hazardous test flights ever attempted.

At first hearing, such work sounds impossibly dangerous, but these tasks can be done without fatalities, day after day, for years. To use the barrage-balloon example, the RAE men were test pilots at heart, and so they knew to approach the problem in slow stages. First they flew against fishing lines hung from parachutes and by stages worked their way up to ramming steel balloon cables, eventually developing explosively powered cutters to break the fouled wires before a wing came off. But it was never safe, as evidenced by the cables that some-times got caught in a propeller.

To tackle such jobs without the keen edge of fear invites disaster. These settings tell us a few things about the right personality for dangerous work. In my informal polls I heard that the best employees in high-hazard jobs are meticulous and confident. They communicate well between all levels. They enjoy life and they have a good sense of humor. Daredevils, or anyone else with notions of going out in a blaze of glory, need not apply. These people are in for the long haul.

For decades ammonium nitrate aroused no fear among those who handled it, which to me explains why disasters with ammonium nitrate have killed many more people than those with nitroglycerine, even though ammonium nitrate is a less powerful explosive, pound for pound, and much more tolerant of careless mistakes. Despite an earlier string of fires and explosions with the stuff, the makers and users of ammonium nitrate did not have to confront its dark side like the makers of nitroglycerine had to, decades earlier. But that all changed in the *annus terribilis* 1947, when four shiploads of what most people thought of as a garden-variety fertilizer blew up in three port cities.

LIKE NOTHING ELSE ON EARTH

It didn't take long for nitroglycerine to win the reputation as a sort of *eau de havoc*. Ascanio Sobrero discovered it in 1847 while a professor of chemistry at the University of Turin. Sobrero kept bottles of glycerine on hand in his lab for use as a skin lotion and experimented one day by dropping a little of it into a mixture of sulfuric and nitric acids. The nitrating reaction yielded a yellowish liquid that, he found quickly, exploded over a flame. The biological effects interested him: a small amount made a dog violently ill. Sobrero could also guess by the way the dog banged its head on the wall that nitroglycerine was giving it a roaring headache, as bad as Sobrero himself had gotten when he tried a drop of the stuff on his own tongue. Sobrero passed his findings on to the scientific community and thereafter tried no more experiments with the stuff. A British doctor discovered it was good for treating chest pains from heart disease.

Other inventors realized nitroglycerine had great potential as an explosive, but none of them could make it go off on command. Sometimes it detonated, but at other times it just caught fire. One of those men was a stooped young Swede inclined to be sickly all his life. Alfred Nobel was born in 1833, the same year his talented but eccentric father, Immanuel, went bankrupt. It was but the start of an up-and-down childhood.

Nobel grew up with explosives. His father's business of bridge building required him to blast obstructions loose with gunpowder, which eventually led Immanuel to the invention and manufacture of antiship mines, then called torpedoes. After a prototype torpedo exploded and wrecked his house in Stockholm, Immanuel moved the family to Finland and then Russia and continued his work at the urging of the czar's naval officers. It culminated in a nest of torpedoes to protect the Fortress Kronstadt, which defended St. Petersburg from enemy warships. But the Crimean War ended in 1856, and the Russians had no further need of torpedoes. So at age sixty, the elder Nobel took his wife and two youngest sons, Emil and Alfred, back to Sweden. They began making and selling Nobel's Blasting Oil out of a shop in Heleneborg, outside Stockholm.

This company would bring Alfred Nobel business fame but also alienate him from his father. Immanuel would come to regard his son as making off with his idea about how to detonate nitroglycerine reliably.

Mixing up a batch of nitroglycerine was comparatively easy, but coming up with a reliable way to set it off was very hard. Alfred, apparently following a line of thought originating with his father, found a way to do so in 1864. Like his other famous invention, dynamite, Nobel's first good detonator seems thunderingly obvious in retrospect, except that nobody outside the family had thought of it before. It was a container of gunpowder, basically a big firecracker, stuck into a container of liquid nitroglycerine and set off with a fuse. The gunpowder's explosion set off the liquid. With that proof of concept, Nobel moved on to a commercially successful detonator. It was a small copper tube filled with fulminate of mercury, still used today in making percussion caps for ammunition.

In October 1864, a few months after Alfred Nobel got his detonator working, the family's tiny nitroglycerine plant blew up. The fire and explosion killed four people, including his younger brother, Emil. It might have gotten Alfred, too, had he not been away talking to a potential investor. Stockholm banned any more dealings with nitroglycerine inside its boundaries, but Alfred had no time for second thoughts. The emotional shock from the explosion gave his father a stroke from which he never recovered, so Alfred and Robert had to support the family. Alfred moved his operations to a barge anchored in Lake Mälaren and got back to work. He made enough caps and nitroglycerine to fill a suitcase, and he hauled his product to market.

Within a year of his brother's death the new business was, well, booming. Alfred Nobel and Company plants were established in Sweden, Germany, and Norway. Blasting Oil was proving itself in mines and tunnels across Europe and America. Though a liquid was not ideal for use in drilled holes, hard-rock men liked the way its explosion shattered rock into small pieces. Gunpowder would split the rock off in great chunks, each chunk needing to be split again.

Nobel came close to losing his business to public outrage over the next two years after a series of blasts from stored nitroglycerine. One of the first happened because a German salesman left an unmarked wooden case of canned nitroglycerine at the Wyoming Hotel in New York, saying he would come back for it later. The porter used the box as a footrest in his shoe-shining business. Three months later, in November 1865, guests drinking in the hotel bar smelled chemical fumes. They traced it to the box, now in a storeroom and leaking pink smoke. They dragged it into the street and left it there. It exploded seconds later, digging a foxhole in the street and injuring eighteen people. And over the next year more accidents happened, with explosions of stored nitroglycerine occurring in Sydney, Australia; at a Wells Fargo freight office in San Francisco; on a steamship in a Panamanian harbor; and near Brussels, Belgium.

Humanity had never encountered anything like the stuff. Here was a liquid that could sit in a closet for months, then start fuming one day and bring down the house ten minutes later. Every container of it was a time bomb answering to its own unreadable clock, or so it seemed.

Nobel and his brother Robert worked harder than ever in the laboratory to conquer the problem. They found the principal cause of its instability was leftover acid from the manufacturing process. This caused the nitroglycerine to degrade with time, making it prone to a runaway reaction as contamination built up. If freed of all acid and stored in a cool and stable place, nitroglycerine would not explode on its own. Proof is that seventy years after Sobrero's pioneering work, chemists came across several glass vials holding his original samples, as good as new. They may be around yet.

This discovery made storage safer, but transportation was still a problem. Other people had tried mixing nitroglycerine with various materials to make it safer, but none of these experiments had proved worthwhile. Nobel knew from his earlier research that the powder he packed his vials in, a whitish clay called kieselguhr, would soak up three times its weight in nitroglycerine. He found that it also made nitroglycerine much safer. *Dynamite*, his name for the mix of kieselguhr clay and nitroglycerine packed in paper tubes, is still in use, although clay is no longer used as an absorbent.

Safety strides in the use and storage of dynamite narrowed the principal danger down to factory workers, such as those who had to tend the equipment that made it. One American company had sixteen nitrators blow up in the decades from 1915 to 1955, costing sixteen lives. But after the invention of a much safer method called continuous nitration, over the following two decades that same company blew up only one nitrator, with no lives lost.

In 1959 there were thirty-four dynamite plants scattered across the United States. A few companies still manufacture nitroglycerine today for medical purposes and as a key ingredient of the smokeless powder used in small arms. But only one plant in North America still makes it for dynamite manufacture, Dyno Nobel's Carthage, Missouri, facility.

I spent a July afternoon there to see the trade of making what the workers called simply "oil," or sometimes "NG." (I never heard them call it "nitro.") The plant is marked only by small white signs on a county road, followed up by stern warning placards at the gate: no smoking, no matches, no cameras, no firearms. Safety director Rick Fethers began our tour by handing me a map of the plant, divided into four evacuation zones. The idea is that an impending fire or explosion will offer enough notice that somebody can post an alarm, which will set off an appropriate number of honks from the plant's warning horn to inform employees which zone has the problem. In that way people are more likely to run from, rather than toward, the danger. Fethers asked if I had any matches or lighters, a notorious source of death and destruction for the old explosives industry. Lammot du Pont had been so deadly serious about stopping employees' smuggling of smoking materials that before each work shift he had the workers strip and walk through water neck-deep before donning work clothes.

I left behind my notepad and pen, along with everything else from my pockets. That meant coins, keys, wallet, cell phone, and anything else that might fall to the floor. The industry knows from experience that a thin film of nitroglycerine over a hard surface, like concrete, is extremely sensitive to dropped objects. Though the workers don't intentionally leave NG lying around in puddles, the unexpected awaits.

We took off in Fethers's 1979 cream-colored Mercedes-Benz, which he swung confidently down the gravel roads of the two-square-mile compound. The Carthage plant makes twenty-four tons of nitro-glycerine a day for blending into various grades of dynamite. Fethers explained as he drove that the mixing buildings and storage magazines at Carthage have been engineered with weak and strong walls to channel a blast away from other buildings. Magazines, mixing houses, and nitrating buildings are isolated from each other according to an official Table of Distances that predicts how far heavy objects might fly through the air after an explosion, from a "donor" building to an "acceptor" building. This reduces the chance that metal shrapnel and burning fragments will set off more explosions. Safety codes prohibit buildings with machinery from holding as much explosive as a magazine because the chance of a detonation is higher if machinery is around.

The Carthage plant learned from its last big blast in 1966, when the plant was under different ownership. Investigation suggested that overheated brakes caused the tires on a trailer truck to catch fire as the truck sat on a loading dock. The heat caused explosions that set off the magazine, which launched more explosives and fiery debris around the rest of the plant. By the time the chain reaction was over, early the next morning, 90 percent of the factory buildings were down. The shattered window glass on the ground reflected the sun that day, and to observers in airplanes it had the look of thousands of tiny triangular lakes.

Fethers drove me to the cluster of white buildings where Dyno Nobel makes nitroglycerine, just up the hill from the crater left by the batch nitrator's explosion earlier in the year. A flashing red light on a rooftop signaled that the nitrating process was under way, so no one was allowed into the processing area without clearance from the control room. For this, Fethers handed me over to operator Jesse Whitesell. Whitesell, tall, friendly, and with a slow and thoughtful way of speaking, called the control room on a handset radio before entering each building, then radioed again as we left each building. In that way the operators could follow what we were doing via closed-circuit cameras, which were installed to, among other reasons, make post-explosion analysis easier.

In the decades of explosives manufacturing worldwide, there have been some plant explosions with no apparent cause, such as a 1971 explosion at Würgendorf, Germany, where an intense investigation turned up nothing. And investigators can't depend on having witnesses to interview. Some blasts, like one in Great Oakley, England, in 1928, removed every trace of the people nearby, as if they had been plucked from the face of the earth. Though such cases were probably the usual runaway reactions, there's always the possibility that something is missing from the short list of things we know to avoid with nitroglycerine—like shock, heat, and acidity.

After the explosion of its batch nitrator in May 2000, the Carthage factory had only one nitrator left—the newer "continuous nitrator"—so we visited that one. Here, nitroglycerine (technically, "ethylene glycol dinitrate") leaves the nitrator as milky-white droplets in a stream of water and flows down a covered trough to another building for separating and washing. On the inside, these buildings have the look of dairy operations: clean concrete floors, whitewashed walls, and stainless-steel pipes and vats. At the last point of the process, fresh nitroglycerine flowed from an elbow-shaped tube and gurgled peacefully down a drainpipe, with a few inches of air gap between. Whitesell said it's safer than confining it in a tube, where sudden compression of the liquid could set it off, perhaps by someone shutting or opening a valve too quickly. Because nitroglycerine is a thick liquid that always holds some trapped air bubbles, if the liquid is compressed quickly the air in the bubbles heats up, and this heat can set off a runaway reaction that detonates an entire batch. It's called adiabatic heating and explains many strange explosions that have worried investigators over the years.

Nitroglycerine loses its liquidity at Dyno Nobel's dynamite-mixing house. Built on a slope at the foot of a hillside, a reassuring distance away from the center of the plant, the mixing house is an aged wooden structure that looks from the outside like something miners might have used to sort ore during the Colorado gold rush. But inside, it was clear that no rocks are going to be sorted or crushed here. Mixing dynamite is an exceedingly gentle operation, using bakerylike mixers to knead nitroglycerine with dry, absorbent materials like sawdust and chemical boosters.

"Smell that?" Fethers asked. The nitroglycerine odor was not strong or unpleasant, but it was unlike anything I'd ever smelled before. The smell reminded me of the disinfectant that hospitals use, with a faint overtone of pool chemicals. After a couple of minutes, I could feel a dull headache coming on. The body absorbs nitroglycerine fumes quickly, which makes the blood vessels expand and causes migrainelike pain. A new worker is apt to be truly miserable at first, but then his body adapts. The headaches will stay away for as long as he remains exposed to nitroglycerine on a daily basis.

The workers in the mixing house were on lunch break, their white cotton gloves left in orderly fashion on the worn wooden treads of the stairs. It was peaceful and quiet here, in a timeless way; safety codes prohibit portable electric devices such as radios or water coolers. Every nail head in the building was covered up. All the carefully taken precautions make for a reasonably safe operation, but it also means that methods have to stay frozen in time. Nitroglycerine is such a harsh mistress that Dyno Nobel can't automate operations here like it can with the less sensitive ammonium nitrate explosives, which it mixes up elsewhere on the factory grounds.

By World War II, people in the explosives business had reached a certain peace with nitroglycerine. Though occasional factory explosions still happen today, the damage is limited because people know to keep their distance. The frightfulness of a mistake has become the ally of safety.

One of the things nitroglycerine safety always had in its favor was that the stuff was known to be hazardous from the outset. While discovered to be good for a few other things, such as treating heart conditions, nitroglycerine retained its "mind share" as a very touchy explosive.

EXPLOSIVES IN A PLAIN BROWN WRAPPER

For a long time this attitude of healthy fear wasn't the case with those who handled ammonium nitrate, which is a highly useful fertilizer with an alter ego. It is so powerful that explosions involving ammonium nitrate have blasted thousands of tons of concrete foundations out of the ground, hurled great hunks of steel nearly three miles, and knocked planes out of the sky. That's because for the longest time, peo-

ple couldn't think of ammonium nitrate as anything but bagged fertilizer, despite growing evidence to the contrary.

If you take care of a lawn, you've probably handled it often. Right
out of the lawn-food bag, it has the look of brown, coarse sand. It absorbs
water easily. It's a salt, in the same broad chemical class as ordinary table
salt, but unlike table salt, ammonium nitrate has oxygen embedded in its
molecules. It became an important fertilizer for a hungry world early in
the twentieth century, after the Germans invented a way to make it
cheaply out of natural gas and nitrogen from the atmosphere. Added to
the soil, ammonium nitrate brings usable nitrogen to crops that would
otherwise have to scrounge for it through slow natural processes.

Starting in late 1945, the United States began making and shipping
up to one million tons a year of ammonium nitrate to Europe, to boost
food production there. The ammonium nitrate came from former ammunition plants, which had been making it during the war for explosives
manufacturing. Mixing nitric acid with ammonium hydroxide yielded a
watery liquid that, when heated, evaporated and left behind white grains
of ammonium nitrate. Workers coated the fertilizer grains with water-
repellent wax and loaded it into hundred-pound bags.

One of the ports receiving the fertilizer was Texas City, Texas.
When boxcars arrived there, stevedores unloaded the bags. Some of the
men said later that some bags were still warm when pulled from the
boxcars, even charred, which might have raised suspicions that the bags
were reacting with something and generating their own heat. The
charring made the bags liable to break and leak fertilizer on the
ground. Rather than waste good fertilizer, the port's workers scooped
up the stuff and sealed it into empty bags, along with spilled grain, and
loaded these bags along with the unbroken ones. The practice appeared
to be good housekeeping, but in fact it was the mixing of a fuel—the
spilled grain—with ammonium nitrate.

The city's wharf area offered three slips, or inlets, for freighters to
pick up their cargo. Each slip was a thousand feet long and wide
enough to serve ships berthed on both sides. One of the slips, the north
one, lay alongside a long building labeled Warehouse O. Tall stacks of
brown-bagged fertilizer lay in and around this warehouse when the

Grandcamp, a French-owned freighter surplused from the American war effort, tied up at the north slip on April 11, 1947. The *Grandcamp* had come to pick up three thousand tons of fertilizer.

The port's origin dated to the 1890s, when three Minnesotans bought ten thousand acres north of Galveston. The new owners dredged a ship channel to Galveston Bay and laid in a railroad, which put them in a good position when the wells at East Texas's Spindletop Dome blew in a few years later. By 1947, Texas City was very much an oil and petro-chemicals town, carving out a niche of prosperity during the postwar recession. Within a mile of Warehouse O lay six processing plants that refined or stored petroleum products, along with two dozen blocks of res-idences. The closest factory—a Monsanto Chemicals plant that made styrene—lay just on the other side of the slip.

By the time the stevedores knocked off work on April 15, the *Grandcamp* had been loaded with twenty-two hundred tons of what the men thought to be plain old fertilizer. The ship also held two hundred tons of peanuts, leaf tobacco, oilfield equipment, giant balls of sisal twine, and sixteen cases of small-arms ammunition, all loaded at other ports. The ship was 437 feet long. All 7,176 tons of the *Grandcamp* would rise up and vanish the next morning, leaving only a vast gouge in the pier.

The stevedores closed the hatch covers that evening and went home. The next morning, on opening up hatch 4 to load the last six hundred tons, a stevedore noticed a smallish fire down in an underpin-ning of wooden timbers installed to keep the bags from getting wet at the bottom of the hold. There's no way to know how the fire started. The cargo could have caught fire spontaneously, but it was most likely the result of a stevedore's discarded cigarette having been dropped into the hold. A fire had been started in that way just two days before, as men working along the docks had routinely ignored the No Smoking signs.

A few pails of drinking water and a handheld extinguisher didn't get the low blue blaze under control. The men started to lay out a fire hose, but a ship's officer, probably Captain Charles de Guillebon, told them to stop because water would damage the cargo. Instead, the offi-cer ordered the men to pipe in steam from the engine room, figuring this would put the fire out by starving it of oxygen.

"Steaming the hold" was a traditional fire-fighting solution for the unusual emergency that had befallen them. Steaming was the worst thing they could have done, short of lighting up a flamethrower and dropping it down on the fertilizer bags before battening down the hatches. The words of John F. Kennedy come to mind, describing an incident during the Cuban missile crisis: "Always there's some son of a bitch who doesn't get the word."

The ones who had the word were the Army Ordnance Bureau's Emergency Export Corporation, which was overseeing the manufacture of the fertilizer, and the U.S. Coast Guard, which had written regulations stating that the chemical was a "dangerous substance." By 1947 the Coast Guard had clear authority to regulate the arrival and loading of such shipments, but it was short of money and hadn't fully taken up the duty. The Coast Guard's nearest port official, at Galveston, said later that he didn't know that Texas City was a major through point for ammonium nitrate. Once the war ended and the army stopped managing shipments of it, the Coast Guard by default allowed stevedores and ships captains to treat ammonium nitrate like any other cargo. The Coast Guard believed that shippers would be self-policing when it came to port safety. It was an odd attitude for an agency specifically charged with preventing port disasters comparable to the 1917 explosion of twenty-five hundred tons of explosives aboard the French freighter *Mont Blanc* in the Halifax, Nova Scotia, harbor.

While ammonium nitrate was not like TNT in its properties, there were more than a few hints of its explosive qualities. There was the odd case of the Badische Anilin und Soda Fabrik, or BASF, factory in Oppau, Germany. This factory made fertilizer, and was the first plant in the world to create ammonia from air and natural gas. Great piles of mixed ammonium nitrate and ammonium sulfate were held in the storage yards in Oppau, where they were exposed to the elements. Rain and snow had formed a hard shell on the piles, and so workmen needing to draw material from them used small explosive charges to blast loose what they needed for the work shift. More than fifteen thousand times the blasting occurred without incident. Then on May 21, 1921, the same technique set off two successive blasts in a forty-five-hundred-ton pile of fertilizer, destroying the factory and digging a crater 450 feet across. That explosion killed 560 people.

The Oppau scenario played out again in 1942 at a fertilizer plant in Tessenderloo, Belgium, when workers using small charges to loosen outdoor piles of potassium chloride detonated 150 tons of ammonium nitrate instead. The plant had used both types of chemicals, and because they looked so similar, the workers had lost track of where the ammonium nitrate had been placed.

Not knowing about these accidents, dock and ship workers in Texas City could not be blamed. The simple fertilizer labels on the bags didn't tell them that ammonium nitrate is an oxidizer; that is, that it would give off oxygen under certain conditions and thus become dangerous. However, both the army and the Coast Guard were aware of this, and had the latter been supervising the operation—rather than leaving that task to the shipyard—the disaster would not have occurred.

A powerful stream of seawater from the fire hose would have saved the port. But instead the men closed the hatch covers, shut the ventilators, and poured on the steam. Down in hold number 4, the cargo started to melt when it hit a temperature of 338°F. The only sign of the complex little chemical reaction was the way that the smoke changed color. Minutes after the men shut the hatch cover, it began to jump up and down, as if something was trying to escape. The heat mounted, and the stevedore company called off its men at 8:30 A.M.

As the fertilizer heated, the ammonium nitrate broke down chemically and approached what one might call a chemical fork in the road. In one reaction, it could produce water vapor and nitrous oxide, called laughing gas. This decidedly unfunny reaction gives off extra heat, so things would go out of control quickly if that reaction took over.

The other possible chemical reaction, which occurs at a lower temperature, produces nitric acid and ammonia. A ship making laughing gas may sound a lot better than one belching fire, nitric acid fumes, and toxic ammonia, but strangely the city was relatively safe as long as the second reaction lasted because it sucked up heat from the surroundings and therefore would keep the burning sacks and wood from making the rest of the cargo too hot. Had the cargo confined itself to making ammonia and acid gas, the *Grandcamp* would have burned out without an explosion.

The closed hatches and the unintentional fuel in the hold directed the ammonium nitrate toward the first chemical reaction and its ever-accelerating heat production. The fuels were the wax coatings on the fertilizer grains, the asphalt-coated paper bags, and the feed grain mixed into bags that workers had diligently refilled and repaired back in Warehouse O.

The first sign of serious trouble was the white nitrous oxide gas that billowed out after the hatch cover blew off. Next was the orange and reddish-brown smoke that some onlookers described as "beautiful to see." The strange sight drew hundreds of spectators to the dock and to the head of the slip. The smoke was nitrogen oxide, the same stuff given off by nitroglycerine when it is fuming and about to explode. The ammonium nitrate was in a full runaway reaction now, needing only to reach the explosive threshold temperature of almost 850°F.

Twenty-seven firefighters from the Texas City Fire Department arrived, along with fire crews from nearby petrochemical plants. The sirens drew still more sight-seers alongside the burning ship. The last photographs taken, just before 9:00 A.M., show the water flashing to steam as it hit the smoking sides of the *Grandcamp*.

At 9:00 A.M. flames rose out of the open hatch. Twelve minutes later, a little more than an hour after the stevedores first discovered the problem in hold number 4, nearly twenty-three hundred tons of ammonium nitrate in the *Grandcamp* exploded in a blast heard 150 miles away. A piece of oil-drilling equipment from the cargo turned up two and half miles away, driven six feet into hard clay. The concussion destroyed two light planes passing a half mile overhead. It killed all the nearby firefighters and half of the 450-person workforce at the Monsanto Chemical Company's plant across the slip. Surprisingly, some of the bystanders within two hundred yards survived, but the blast ruptured their eardrums and slimed them with a brown coat of oil and water. One man standing one hundred feet away woke up almost a mile off, apparently carried there by a miniature tidal wave.

Worse still, the blast caught up the freighter *High Flyer* moored in the main slip to the south, ripped loose her mooring lines, and rammed her sideways into the *Wilson B. Keene* on the opposite side. The steel plates in crumpling locked the two ships together, like two semitrailer trucks might entangle in a high-speed collision. But neither ship was yet on fire. The *High Flyer* had 961 tons of ammonium nitrate on board, along with two thousand tons of sulfur.

Tugboats with volunteer crews arrived from Galveston that night and set out lines to pull the *High Flyer* away. But the rubble from the blast blocked their efforts, and they called for more assistance. The ship and cargo were on fire by then. Men with acetylene torches cut the *High Flyer* loose before midnight, working in the stench of burning sulfur and fertilizer, but the tugs couldn't move the ship. The stern anchor had fallen and lodged in the harbor bottom. Shortly after midnight the anchor came loose and the tugs got under way. The *High Flyer* exploded minutes later. The final tally at Texas City was at least 468 people killed, 100 more gone without any trace, and 2,000 injured.

Six weeks later, in the harbor at Brest, France, a cargo of ammonium nitrate caught fire aboard the *Ocean Liberty*. Exactly as with the *Grandcamp*, the crew sealed the hatches and flooded the space with steam, trying to starve the fire. The explosion killed twenty-one people, and a third blast followed later that year when another freighter loaded with ammonium nitrate caught fire in port on the Black Sea. Finally, the explosive properties of ammonium nitrate when mixed with fuel had won the full attention of the chemical industry, not only as a safety matter but as an explosive that would be cheaper to manufacture than dynamite.

As chemists perfected ammonium nitrate–fuel oil explosives, or ANFO, they found that pure ammonium nitrate is reasonably safe from explosion until mixed with a carbon fuel and ignited or detonated in a tightly confined space. Ammonium nitrate itself doesn't burn, technically speaking, but it breaks down upon heating, and the byproducts of its breakdown do burn. It's complex enough that even today firefighters cannot know for sure what a burning warehouse full of fertilizer will do next. But they remember from 1947 what it might do.

Ammonium nitrate is such a commonly used chemical that we

can only hope the fear and respect it inspires among its handlers will last. The handling of some chemicals becomes more dangerous with time because people have lost their fear. Gasoline has gone through this cycle: from fear (at the outset of the internal-combustion era) to respect to complacency. Complacency may explain how on September 30, 1999, a factory in Japan, of all places, could have gotten so careless with uranium oxide as to create an accidental nuclear reaction by shortcutting the processing steps. It happened at the JCO Tokai Works Conversion Test Facility in Tokaimura, where workers thought they could save a little effort and time by filling buckets of acid with dissolved uranium oxide and carrying them over to dump in a precipitating tank. On the day of the accident one too many bucketfuls dumped into the precipitator brought together enough uranium to make a temporary low-grade reactor. The reaction bathed three workers in high-energy radiation for a fraction of a second, then settled down into a critical mass of lower output. The rays of gamma and neutron radiation injured two workers and killed another.

LIVE-WIRE ACT

One class of workers who know the benefits of fear are helicopter crews that maintain live high-voltage power lines. In this next example, we can see what people are capable of when they face up to what could be a very hazardous situation. High-voltage transmission lines carry 230,000 volts or more through wires suspended high above the trees from tall metal towers. Sooner or later, such lines need attention. Maintenance can include changing out hardware on the lines, splicing in new cable, or adding fiber optic cable. A lineman can do all these jobs from a helicopter, while the juice is still on. Only a handful of private contractors in the United States do this type of work. Public utility companies do the rest with their own people and equipment.

Hovering a turbine-powered helicopter loaded with jet fuel, just one foot away from an energized high-voltage power line, with the rotors overlapping the cable sounds insane, especially to pilots who have been taught that all power lines are poison. But done right, this live-wire act is as safe as working out of a bucket truck that raises linemen to the tops of the poles. According to USA Airmobile Incorporated of Fort Lauderdale, Florida, its crews have logged more than fifteen thousand flight hours of this work without a power-line fatality.

First, a word about their workaday world. Big transmission lines usually have three "conductors," one for each phase of the three phases of electrical power. Each conductor is made up of several wires spaced a foot or so apart. The conductors run in parallel directions and are separated by twenty feet or so. These wires are commonly aluminum with a steel core, an inch or two in diameter.

In the old days, linemen climbed up the towers and clambered across to the conductors, but that required the power company to shut the power off beforehand. Today's power companies prefer not to shut down their transmission lines for routine maintenance. Closing down a five-hundred-thousand-volt line, for example, can cost $50,000 an hour. And some utilities couldn't shut down the big lines even if they wanted to, during times when they have no transmission capacity to spare. That's where helicopters come in, flitting from span to span like high-powered hummingbirds.

Typical work involves taking equipment on or off the power lines or replacing a section of frayed wire and splicing in new wire. A line-man does much of the work while sitting on a temporary aluminum platform mounted on the helicopter landing skids. The platform is a stout aluminum plate about the dimensions of a long tabletop that sits firmly on the helicopter's skid tubes and extends beyond them on either side. Because there is a possibility the pilot would have to jetti-son the platform if it hung up on a wire, the lineman always attaches his safety harness to the helicopter's frame. For other tasks the crewman might have to hang fifty feet or more below the helicopter at the end of a string of "hot sticks," which are rigid, nonconducting fiberglass rods. Since hot sticks do not conduct electricity, the helicopter can use the string of hot sticks to drop off a lineman at places where the pilot could not otherwise approach without risking a crash.

Airmobile relies heavily on Bell JetRanger helicopters that they rent from commercial operators. The JetRanger is a single-engine turbine-powered model of high reliability. The two blades of the main rotor are stoutly built. The owner of Airmobile, Michael Kurtgis, says on three occasions that he knows about, the rotor blades in contacting a power line broke through the cable, allowing the helicopter to land safely. Company procedures limit the fuel to twenty gallons, and pilots need to refuel before the level drops to ten gallons, so flights are not long. Airmobile pilots routinely run the Allison gas turbine at 100 percent of its rated power while hovering, backing off a bit each time they change positions.

Airmobile's chief pilot is Doug Lane. He flew helicopter gunships for a year in Vietnam and was shot down twice; on the way to his job at Airmobile he flew freight to mountaintops, did traffic reporting, and shuttled oilfield workers to platforms in the Gulf of Mexico. According to Lane, in addition to learning new flight regulations and electricity and company procedures, an incoming pilot needs to break through some mental barriers. One of the most difficult for newcomers is the nerve it takes to bring the helicopter close enough. If the pilot keeps the machine too far away, the linemen have to stretch to reach the work, which is exhausting at best and impossible at worst, considering the heavy pulleys and clamps the work requires.

Airmobile also trains its pilots in teamwork skills. Teamwork addresses a longtime problem in flying, one reflected in the terminology of the official logbooks that pilots use to record their flight hours. The FAA logbook shows how many hours the flyer has spent as "pilot in command." When I first flew I loved the sound of that phrase: I was the one solely responsible for the safety of everyone on board, even if it was only me and a friend or two. But experience has shown that too often the Pilot Who Would Be King model gives us pilots who ignore, or even intimidate and silence, people around them who could be of great assistance in an emergency. Such people might include a copilot, an engineer, or an air traffic controller. In the case of power line work, this help would probably come from the lineman, who is an electrician experienced in transmission lines.

Of what help could a nonpilot be? Some of the work involves the pilot flying the helicopter slowly alongside a power line, while the lineman checks for damage to insulators and other gear with binoculars or a videocamera. After a long day of this type of work the pilot may lapse into a sort of highway hypnosis, wherein he watches only the line. This is potentially very dangerous because of the risk that the helicopter will collide with a set of wires crossing at a right angle to the ones the pilot has been following. The lineman's judgment is important here: to him, a power line is not a mishmash of poles, towers, and cables but a system that changes its appearance as the power line approaches a substation or a crossing. Even without seeing the interfering wires coming up—and these can be very difficult to spot when a helicopter is flying as low as these have to—the lineman is more likely than the pilot to notice the subtle signs of changes ahead.

Teamwork is equally important on the ground, when the crew is working out a plan of attack before flight. Airmobile policy gives the pilot and lineman equal rights to question a procedure that looks unduly risky to either one. This might be a disagreement on whether to take the helicopter near sloping wires on a hillside, risking a "wire strike" with the rotors, or whether to go into a tight place at a time when wind conditions are on the edge of gusty. Skilled pilots can handle thirty-mile-per-hour winds if the air is flowing in a steady fashion, but much less than that if the air is turbulent. In case of such a disagreement, pilot and lineman would have to consult management or strategize a safer approach.

Another way that these crews embrace danger is to connect themselves electrically to the power line while the work is under way. To see this I dropped in on a crew employed by Haverfield Inc., working out of a pasture set in the Appalachian Mountains of Pennsylvania. Pilot Mark Campolong and foreman Ken Black started the workday, as they do every day, with a tailgate safety meeting. Such tailgate briefings go by different names, but all have the same essentials: They occur at the start of a shift, they're held on the worksite, they're informal, and they last maybe fifteen minutes. (In fact, an investigative panel after the *Mars Climate Orbiter* spacecraft loss said such daily gatherings, which it called "stand-up, tag-up" meetings, could have helped avoid the loss of that spacecraft.) The meeting that I attended included lineman Jeff

Pigott, dataman Craig McCleaf, and a crew chief from Pennsylvania Power and Light (PPL), the owners of the line.

SHALL WE GATHER AT THE TAILGATE?

Though the same crew had been working from the south end to the north end of the thirty-four-mile Montour-Columbia-Frackville transmission line all summer, taking old hardware off the twin wires of each conductor and bolting on new equipment, and had held such a safety meeting every day, Black went over the mission basics once again. He pulled out a dogeared wallet card listing the minimum distances the helicopter needs to keep between wires, or between a wire and trees, to prevent a short circuit from passing through the helicopter. "Whatever distance it says, we double it," Campolong said. It was also noted that the three conductors were spaced far enough apart that the helicopter could hover alongside the center conductor. And, at Haverfield's request, PPL had set the line to "manual reclosure." This way, if a person got electrocuted it would happen only once, rather than three successive times, as automatic reclosers tried to restore the line to service. Such is the edge of survival in this business.

I asked to go on a sortie with Campolong, who despite the death-defying sound of the job has the quiet demeanor of a Sunday school teacher. The men checked over my borrowed conductive clothing and flight helmet, and we were good to go. The Hughes MD-500 helicopter rose briskly, turned, and headed for a span of power line a half mile to the south. Campolong matched altitude and approached the wire with seasoned confidence, coming to a precise hover two feet away. Jeff Pigott stretched out a metal rod and struck a foot-long arc with the power line, then clamped on a temporary cable that would maintain a good electrical connection between the helicopter's frame and the line. This is called bonding to the wire. The reason for the arc is that the helicopter has a lower electrical potential than the power line, and current wants to jump across the air gap to equalize them, something like the way that water seeks its own level.

That's why the crew members can sit on metal platforms and ride in a metal helicopter: for this brief time, all the people in the helicopter are wearing the full voltage of the transmission line and are safer than they would be by trying to insulate themselves from it somehow. From here a passenger can admire the pretty trees, but—if a branch ever came close enough—to reach out and touch this little bit of nature would kill everyone on board instantly, as the full power-line current would take a shortcut through the helicopter's frame to the ground.

"You've got two hundred thirty thousand volts flowing around you now," Campolong said.

Campolong stabilized the helicopter to put the live wires about level with Pigott's stomach, and a few inches above his legs. Campolong kept a steady gaze in Pigott's direction, sparing only quick glances at his instrument panel and the sky. His skill made it all look easy, but holding a helicopter steady under breezy conditions with only a few inches' margin of error takes intense concentration. One pilot compared it to riding a unicycle on a walkway belt (like big airports have) while somebody randomly changes the speed of the walkway. Even a few minutes of such precision hovering is enough to exhaust a pilot trying it for the first time.

After six minutes or so Pigott bolted the last of the new braces on this span of line, and Campolong banked away. "Piece of cake," he observed as we flew back to their landing zone.

Even after hearing about the safety precautions that workers like Campolong adhere to each day, you could well ask why anyone would accept these kind of risks when other, safer jobs are available. One reason I'd offer, from my days helping my two brothers blast out an excavation site with high explosives, is that it clears the mind. I found it refreshing to work with something so devoid of foolishness. No one would be around to correct a careless mistake, and there would be no second chances. I also encountered this no-nonsense attitude once while spending time on a drill ship in the Gulf of Mexico. Shortly after my helicopter landed, the medic on board insisted I remove my wedding ring. I desisted. I had never taken it off, even once, since getting married sixteen years before. But the medic persisted. Jewelry was only sentiment, after all, and he had seen too many cases of power tools flaying a finger to the bone after catching a ring, or of burns sustained after a ring made a short circuit in electrical gear. He clinched his argument by saying that we were several hours away from the nearest hospital, and he would not be able to save a finger given the limited onboard medical facilities. I took the ring off.

TOUGH CHOICES

Sometimes the best our technology can offer is the lesser of two risky options. The crew of the *Petroleine*, a sixty-three-hundred-ton British ship crossing the Atlantic in a loose convoy bound for New York, met such a situation in March 1918. When the ship was a day out to sea, a German submarine surfaced nearby and began firing its deck gun.

With full knowledge and approval of the captain, Chief Engineer Towns and his three assistants screwed down the steam safety valves on the boiler and began chucking coal into the furnaces with a will. In naval lingo, they were "forcing the boiler." As German shells fell around the ship, some landing in the sea near enough to drench the gun crew at the stern, the redlined triple expansion engine pushed the ship to a new speed record, faster even than during its sea trials ten years earlier. After an hour-long chase, the *Petroleine* pulled away.

Was it dangerous? Certainly. Many boilers have exploded under such abuse, but in this case, valor was the better part of discretion. The *Petroleine* was a tanker returning empty to the United States, but its big compartments still held enough explosive power in the form of gasoline vapor to blow the vessel apart if hit by a shell. The engineer officers got two months' extra pay for their desperate hour in the stokehold.

We've seen how the safety attitude has been built up through long and hard experience by those who know they are dealing with hazardous situations. Most work settings are not so obviously dangerous, yet they also have their share of devastating mistakes. So we turn now to ordinary human errors with extraordinary results.

10: THAT HUMAN TOUCH

HOW LITTLE ERRORS MAKE
BIG ACCIDENTS

This December night was turning into a nightmare. Surely something had broken and jammed the police van's throttle open, so if Officer Thomas Sawina stepped on the brake pedal as hard as he possibly could, maybe the van would stop before anybody on the holiday parade route got hurt. According to the witnesses who saw the police van speed by with its squealing tires, Sawina's expression was one of sheer terror. His arms were rigid, his eyes were "big as saucers," and his body was forced back against the seat, almost as if someone had him gripped from behind. But what had the officer in its grip was not a person.

People screamed and scattered as the van accelerated toward the office building, bumping over the curb, its rear end fishtailing. One man grabbed his boy and threw him off to the side, out of the van's path. The van caught a middle-aged woman trying to protect her children and ran her down. It scooped up two strollers with children and hurled one toddler through a plate-glass window. Scott Gerlicher, just to the north with his wife and two daughters, had watched the van accelerate and ram the squad car and then the Northern States Power Company building. The crash was over, Gerlicher thought, but then he saw the van bounce off the wall and surge forward again, still at full throttle. Now it was coming for his family.

This kind of power, so easily taken for granted now, lets minor errors wreak major havoc in ways undreamed of during the horse and buggy days. Such errors can be the same kind you've made often: aiming for one button and hitting another, neglecting to do something you had meant to, taking an impromptu shortcut. And the havoc is not all caused by vehicles, reactors, or exploding oil rigs, either. Analysts say that simple human errors in the health-care field may be killing as many as a hundred thousand Americans each year. In such circumstances, and in just a moment, a full career of good work can be wiped away. As a quality-control supervisor for a NASA contractor reminded his workers during the Apollo days, it takes just one "Oh, shit!" to replace a thousand well-done tasks.

FOR THE KIDS

Every year in November and December, the Minneapolis Downtown Association sponsors a free parade featuring floats aglow with thousands of tiny low-voltage lights. The hour-long "Holidazzle Parade" starts on an evening in late November and runs for thirty-two nights. It draws thousands of people, many of them children, to the shops and boutiques on Nicollet Mall.

On the afternoon of December 4, 1998, Kyle Loven, aged thirty-one, was not looking forward to joining relatives at the parade that night. He worked in downtown Minneapolis and so knew the area well. It was a Friday and the weather was warm for the season, so the parade route would be more congested than usual. Loven did not consider downtown

Minneapolis safe at night for a group with small children. Loven found some reassurance from the thought that many Minneapolis police officers would be on hand, keeping the peace.

Five people joined Loven that night: his wife, his one-year-old daughter, his mother-in-law, and a young niece and nephew. The Loven family found a wide-open spot at a brownstone office building on the west side of the street and near the intersection of Fifth Street (see Figure 10). At street level it had large plate-glass display windows that Northern States Power Company (NSP) used to promote its gas and electric services. The building's front also had a low stone wall that people could stand or sit on and an awning that attracted families as a light drizzle began to fall.

With the parade about fifteen minutes away, four drunk men heading south set off a ripple of concern in the crowd at this area along Nicollet Mall. The four men were falling over each other in a chain reaction of drunken hilarity. It left two of them sprawled on the pavement. People called out to policemen nearby. Lieutenant Rick Thomas pulled up in his squad car, parking opposite the NSP building and radioed for the police van called "Detox One" to haul the drunks off the parade route.

An Econoline van with police markings approached at a slow, careful pace from the south, its emergency lights flashing. The driver of Detox One, Patrick Kiely, and the paperwork man, Thomas Sawina, were both off-duty police officers. Kiely and Sawina had been working together off and on since 1990 in a well-settled routine: Kiely drove and Sawina handled the paperwork for each pickup and delivery of drunks to their homes, a detox shelter, or to a medical treatment facility. Eight years before, Tom Sawina had helped set up the detox van program, a federally funded program that tried to accommodate both the neighborhoods' desire to clean up their streets and the needs of down-and-out men. "Tom took care of them, at least for the moment," one officer told me.

Sawina had already put in a thousand hours of detox work in the first eleven months of the year, in addition to his regular daytime job at the Seventh Precinct. He rode the streets three or four nights a week, and

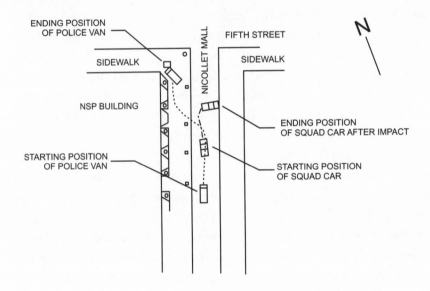

PROBABLE SEQUENCE

1. Police officer attempts to move police van but (federal reports say) presses gas pedal instead of brake when shifting to "drive."

2. Van at full throttle strikes squad car and hits NSP building.

3. Van scrapes along side of building, killing two paradegoers.

4. Pillar at corner of building stops van from continuing.

Adapted from accident police reports

he also prepared duty schedules for the detox teams and the van. As Lieutenant Thomas signaled the van to stop and pick the men up, Scott and Cari Gerlicher looked on, joking that the drunken men must have come to see the parade. They recognized the officers in the detox van; Scott Gerlicher was a police lieutenant himself and had been Tom Sawina's supervisor at the Seventh Precinct at one time; his wife was a sergeant and worked as an investigator. The Gerlichers stood a few feet from the Loven group, who were sitting or standing on a ledge.

Kiely got out of the driver's side of the van to help load up the first man. After eight years' experience, this was routine to the officers of the detox squad. The officers knew most of the drunks by face or name; one of the two men here lying on the wide sidewalk this night went by the first name of Lawrence. The same van patrol had picked him up twice in October for drunkenness. They set the man down on the floor, faceup. Lawrence's buddy was down on the east side of the street, near the squad car, so the officers would have to pick him up, too.

Sawina, standing by, saw that the officers would have to drag the second man in his slippery, wet clothes twenty feet or more, because Thomas's squad car blocked the detox van from getting any closer. Considerately, Sawina walked over, got into Thomas's squad car, and drove it forward five or six feet to get it out of the way. He went back to the van and stepped up into the seat to bring the detox unit closer.

Something subtle but important had happened; the risk of error had gone up. According to the federal investigation afterward, the long pattern of Sawina's detox duty was broken; now he was at the wheel of a van that he did not drive often. Sawina's driving habits were adapted to the controls of his other cars, and most recently to the Crown Victoria he had just moved, which was like the one he drove when on duty. The Crown Victoria's gas and brake pedals lay farther to the right than the van's.

The parade would be starting down the route in minutes; the police needed to clear the area; and Sawina's fellow officers were trying to drag a drunk in the drizzle. It was, apparently, a time to move quickly, even with large numbers of people standing nearby. So quickly that (according to witnesses) he started to move the van with his left leg still hanging in the footwell of the door, his safety belt left unbuckled. Sawina aimed his foot at the brake and threw the shift lever into drive.

Tom Sawina is the kind of man you'd be happy to have as a friend and neighbor. He's a veteran of the Vietnam War and had spent the last twenty-three years in police work. In 1986, at a site just a block away, he had been shot on duty; a bulletproof vest saved him then. His day-time job at the precinct let him do what he liked best, which was to spend his days building a rapport with the lower-income community. He was close to retirement, but had told friends he planned to keep working to maintain a good living for his family, which included two young children. And it was a nice income, reaching $100,000 a year with regular and part-time work.

Even with all his long hours, Sawina made time to exercise before shifts, riding a stationary bike, lifting weights, or running the treadmill. Kiely considered Sawina a health nut, so conscious of fatty foods that he would discard the skin from his chicken at mealtime. Holding back the years at age forty-nine, Sawina had recently gone for a hair transplant. Life was good for Tom Sawina until he put the detox van into drive, shortly before 6:20 P.M. that wet and cold evening.

Kiely glanced up at a loud sound from the engine, and heard a *clunk* as the transmission engaged. The engine noise climbed quickly to a roar, the air cleaner sucked air and made the disturbing hollow sound called "induction noise." The tires squealed and found a grip, and the van headed north at full throttle, its rear doors still open. The detox van rammed the right rear of Lieutenant Thomas's squad car with a great crash, spinning the car in a quarter turn and sending it skidding toward the sidewalk opposite the NSP building. The crash threw Detox One's rear doors shut and deflected the vehicle to the northwest, aiming it like a big white cannonball at the families standing at the NSP building.

The van caught Denise Keenan trying to protect her children and ran her down, ripping her Old Navy jacket. The front fender scooped up two strollers: one with Blake McCarty, Keenan's grandnephew, and the other a blue double-seater holding Keenan's own two children.

The van crashed through the entire Loven group, all except Kyle, who had gone across Fifth Street a few minutes earlier at his wife's request to retrieve a blanket from the car. Detox One hurled his daughter through a plate-glass window, leaving her with internal injuries and two broken legs. It broke his mother-in-law's leg. His niece would lose one arm below the elbow. It left his wife with crushing injuries to her legs and a broken pelvis and collarbone.

Its rear tires each leaving five little parallel stripes of rubber as they spun and smoked, the van ground its way north along the brownstone front of the building, leaving a headlight and piles of shattered glass behind. Blake McCarty's stroller was crumpled by now, but the other was still upright and holding two children. The strange sight of the blue Keenan stroller "jumping and bumping" at the front bumper and threatening to go down or be smashed against the wall, captured the complete attention of Cari Gerlicher. The stroller was sideways to the bumper, its little rubber rear wheels skidding along the wet stone of the sidewalk. Cari had herself and her family to worry about, but she grabbed for the stroller anyway as the van approached and as her daughter Hanna grabbed her leg. It was a breathtakingly brave thing to do, and the city would eventually award her its Medal of Honor for this act.

Cari Gerlicher couldn't pull the stroller free but yanked it around so that it was rolling in the same direction as the van and so less likely to overturn. Now the van bounced off an obstacle and knocked her down. The left front wheel rolled up her leg and onto her pelvis, trapping her daughter's legs at the same time. As Detox One pressed forward and Cari screamed for somebody to make it stop, and as Scott Gerlicher tried to hold the machine back from his wife and daughter by his own strength, the long nightmarish sequence ended. A pillar and doorway at the corner of the NSP building had finally trapped the van.

Some witnesses said the horror lasted for more than a minute, but investigators reconstructed the entire event as lasting about six seconds. The time dilation illusion is common among people under extreme stress; when I was a passenger in a car crash at age fifteen I saw pieces of glass flying through the air in distinct, slow motion.

The trail of destruction, from first tire mark to the gray bashed-in front fender, covered 88.857 feet. Pushed beyond the fender were Blake McCarty's stroller, and the one holding Denise Keenan's children. The first was crumpled, the second fully intact. Kyle Loven sprinted across the street, looking for his family in the blue smoke and confusion. The sidewalk was littered with coats, scarves, and a small pile of hamburgers and french fries. Stepping around Denise Keenan's body, Loven found his wife and daughter in one of the NSP display cases. On the back wall of the display was a full-size photo mural of a silhouetted man lazing in a hammock, on a tropical beach at sundown. Kelsey Loven lay in her red snowsuit at the foot of this mural, faceup and covered with broken glass. The display case advertised NSP's automatic payment plan, and big white letters on the mural read, "Relax! Your bills are being paid."

The stone pillar at the corner of the NSP building had saved the Gerlicher family and untold numbers of other people in the intersection just beyond. That's because sudden-acceleration disasters tend to go on and on until something big gets in the way. In April 1992, a car in New York City left its parking spot to accelerate through two intersections and into Washington Square Park, blowing out a tire as it hit the curb. It jumped into the air, smashed a water fountain, and raced down a walkway, veering from side to side, crashing through crowded benches. The car came to a stop with one person on the hood and several more trapped under the wheels. The toll was five dead and twenty-six injured. A crowd of very angry people gathered at the car, expecting to find someone drunk or demented at the wheel, but found only a perturbed elderly woman who said she was sure she had been stepping hard on the brake.

Though pinned against the NSP building, the police van still spun its rear wheels. The rear end slid from side to side as if the vehicle was seeking a way to bring the whole building down. Scott Gerlicher looked in at the driver's side, where Sawina sat transfixed amid the deflated air bag. The bag had spared him from all injury except a bruised finger and a facial abrasion. Shouting over the engine's roar, Gerlicher told Sawina to turn the ignition off. Sawina did so, and Gerlicher organized a group to pull the van off his wife and daughter. His wife came free from underneath but as the group rolled the van back, it dragged Hanna with it. The tires were sticky with melted rubber. They lifted once more, then pulled Hanna free.

Scott Gerlicher saw his wife would live and, with her urging, gathered his daughter up, fearing the worst. Instinctively he ran for his workplace, the Seventh Precinct police station, just around the corner.

Gerlicher looked over to see someone else running from the scene with the same idea, also holding a small figure: it was Katie McCarty with her son Blake from the crumpled stroller. She was calling for someone to help her baby. Gerlicher stopped at the precinct doors, called over an officer, and they drove Hanna at Code Three speeds to the county medical center. The first to arrive, Gerlicher warned the staff that many others would be arriving from the scene.

In minutes, police cars rolling along Nicollet Mall announced by loudspeaker the cancellation of the evening's parade. As investigators mapped out the scene, the detox van's emergency lights turned and flashed. Its police radio played on, narrating the disaster's aftermath. In all, hospitals received nine wounded parade goers. Denise Keenan had died on the scene, and Blake McCarty died at the hospital that night. Finally, battery drained, the emergency lights and radio on the van winked out.

The first news bulletins suggested that maybe one of the drunken men had grabbed the wheel and gone crazy with the van. That unfounded rumor was replaced by opinions from police chief Robert Olson that something mechanical must have failed, jamming the throttle open. "Nothing else makes sense," he told the press.

JUST ANOTHER SAI

Bob Young didn't believe it. Assigned by his boss at the National Highway Transportation Safety Administration (NHTSA) to look into the case after a request from the Minnesota State Patrol, Young knew the pattern. He had been investigating sudden acceleration incidents for more than a decade, first for the General Adjusting Bureau, then, beginning in 1987, for the NHTSA's Office of Defects Investigation. Cases of cars taking off at full throttle from a slow speed or a full stop had first been reported in the 1930s, rising significantly with the introduction of automatic transmissions. The issue didn't capture public attention until complaints about the Audi 5000 sedan made the headlines in February 1986, later generating a *60 Minutes* episode.

Though the drivers and their attorneys protested vigorously that the throttles had opened by themselves and brakes didn't stop their careening cars, the NHTSA in a definitive 1989 report known as the Silver Book stated that no known mechanical defect could explain how both brake and throttle could fail simultaneously and later pop back to normal, leaving no evidence behind. In nearly all cases of sudden acceleration, NHTSA found, only human error could explain it. The drivers must have pushed the gas pedal instead of the brake pedal.

According to *Road & Track*, the Audi 5000 cars built in 1984 and afterward had minor idling problems that sometimes caused the engine to rev unexpectedly, though not to high power. According to those who study the intricate patterns of human error, the sudden RPM jump helps explain the higher incidence of sudden acceleration with this model, because drivers who had been startled by the rise in engine RPM had tried to hit the brake and missed. An experiment by *Road & Track* had volunteers drive a car around a course with an experimenter in the right seat, equipped with a secret control to rev up the engine. The experiment showed that unexpected, abrupt engine revving could startle some drivers into stepping hard on the wrong pedal. While feet and legs are much stronger than hands and arms, the feet have worse motor skills, particularly in a high-fear situation.

And so, based on the story revealed by diagrams of the crash scene and interviews collected by the state police, and reinforced by his visit to Minneapolis in December, Bob Young saw a simple case of what he calls the "driveway maneuver," the kind of quick, heedless thing many of us do when moving a car a very short distance. Although statistics say that the average driver would have to cruise the roads for hundreds of years before experiencing sudden acceleration, across the entire country it happens often enough that the NHTSA has been able to identify several risk factors most likely to cause pedal error. The circumstances of the Holidazzle crash fit three of the risk factors, namely, an automatic transmission, a driver who did not regularly use the vehicle, and a driver not positioned properly in the seat. Only one fact didn't fit: the van had a safety device to prevent the driver from shifting from park to forward without his foot pressing the brake pedal. So how had the accident happened?

The safety device in question is a "shift lock." I never knew such a gadget existed until my sister-in-law's van broke down in our driveway four years ago. The engine ran well enough but the gear selector lever wouldn't budge, no matter how hard any of us heaved on it. After a tow truck carried the van off for repairs, she found out that the shift lock had failed and had been keeping the lever from moving. A shift lock simply prevents the driver from moving the automatic-transmission gearshift from park to any other position unless she puts her foot on the brake pedal first. Audi during its model 5000 crisis had invented the first shift lock. Later all manufacturers of automatic transmission vehicles had followed with similar devices.

My sister-in-law's broken shift lock was easily fixed, and it hadn't caused a crash; it had only blocked our driveway for a short time. This kind of thing is called a failsafe mode: a device failed but in a safe way, and it also let the operator know that it had broken. Would that all machines failed so benignly, including shift locks under different conditions than in our driveway; conditions like at Holidazzle.

Young took a flight out to Minnesota in December to examine the van in a dark corner of the police garage in Brooklyn Park, north of Minneapolis. Detox One had been stored in exactly the same condition as the wrecker had deposited it. Young had the mechanics charge up the battery so he could start the engine, switching off the emergency lights so the van could idle without the battery running down. Working with the state patrol, Young examined the throttle system and the brakes, from pedal to linings. He tested the engine for RPMs at idle and at full throttle. And he checked for how much force on the brake pedal it would take for a driver to hold the van against a full throttle.

Since this isn't the kind of stunt most of us ever try, it may surprise you to learn that the "brake pedal force" needed to hold that van in check, against a Triton V-8 engine at full throttle, took just fifty pounds of foot pressure from the driver. It takes more oomph than that to get off the couch after the football game. Young says that 99 percent of all drivers are strong enough to stop their cars while at full throttle with the brake pedal. Sometimes Young takes a driver out to a parking lot to demonstrate the power of braking. The driver holds the brake pedal with his left foot and floors the gas pedal with the other foot. It's an unnerving experience, but the brakes hold the car every time.

Young measured the pedal layout in all other cars that Sawina usually drove. All their gas pedals lay farther to the right than the F-250 van. And the F-250's brake and gas pedals were less than two inches apart, so there was little room for error. According to Young, if Sawina tried to place his right foot where the Crown Victoria's brake pedal normally would have been, it could have come down on both the van's brake and gas pedal at the same time, or if he was off by a little bit more, his foot would have landed on the van's gas pedal alone.

Young's official report for the NHTSA was released on January 12, 1999: the van's mechanical systems all checked out, so Sawina's right foot must have been on both the brake pedal and the gas pedal simultaneously. Otherwise the shift lock would have blocked him from shifting into drive. Then, perhaps, his foot slipped off the brake. To Young, there was no other possible explanation for why the shift lock failed to prevent the crash at Holidazzle.

THE WIGWAG

A month after his report, Young visited Minnesota again to speak at a hotel in a northwest suburb of the Twin Cities. The subject of his talk at this crash-reconstruction seminar for investigators was sudden acceleration. After three hours, when Young reached the subject of automatic shift locks and how they work, Don Marose, a state trooper, raised his hand. Marose said that he had been able to shift a squad car into drive without stepping on the brake, shift lock or no. It happened, he added, while the emergency lights were on. Other troopers joshed and jeered, but Young, already suffering from a three-week-long flu episode, felt suddenly worse. He had left something out in his earlier investigation.

"I hadn't had [the lights] on when I checked out the van," Young recalled later, "though I knew they were on at the time of the crash. I began thinking of getting another job." Young suggested they all take a break and go out to the parking lot, where many squad cars were parked.

Sergeant Chuck Walerius went first, starting his Crown Victoria and flicking on the emergency lights. Feet off the brake, he shifted it into drive as easily as if the shift lock had gone away. The squad car jumped forward under the idling engine. Within the hour, Young verified that Detox One of Holidazzle notoriety had the same problem. Something subtle had disengaged the shift lock, which was electrically operated.

Back in the Washington area, Young visited the Montgomery County police garage to ask the mechanics about modifications they made to police cars. One mechanic helpfully showed him a small black device under a dashboard, added earlier that day. Sometimes called a "wigwag," it caused the rear brake lights to flash alternately, left and right, whenever the driver turned on the police light bar on the roof. Police departments felt that flashing brake lights would draw a little extra attention from drivers in cars approaching from the rear. Several manufacturers sold them.

Young went back to the Twin Cities to find that mechanics at the Minneapolis Public Works Department had been adding a similar but homemade device for years now, housing the electronics in a piece of plastic pipe. It changed a circuit numbered "511" on the Ford wiring diagrams. Circuit 511, as it happens, not only energizes the brake lights, it controls the electric shift lock. All told, tens of thousands of police vehicles across the country were cruising the streets with some form of wigwag, riding like a parasite on circuit 511.

This device had been added to countless vehicles despite the warning in Ford's manual for police vehicles: "Do not make electrical connections to vehicle electrical systems not specifically designed for after-market equipment installations. . . . As an example, connection of after-market electrical equipment into the brake light circuit or any other circuit which is connected to the PCM, anti-lock brake computer, air bag system or any other vehicle system will cause vehicle malfunction."

NHTSA's Vehicle Research and Test Center found that the homemade relay made the brake lights go on for 390 milliseconds, then go off for 516 milliseconds, then repeated the cycle. Each time the brake lights came on, the shift lock went off. Bob Young wrote an urgent supplement to the first report. It concluded that Sawina had not pushed both pedals simultaneously. The combination of Sawina's foot on the gas pedal and the deactivation of the shift lock by the wired-in wigwag had caused the crash.

Sawina maintains to this day that he was pressing the brake. An expert witness for his side paints the following scenario: this model of van had no heating for the throttle plate in the engine's fuel system, so it would have been subject to carburetor icing. That evening at Holidazzle, the witness says, the van's throttle plate had started to stick shut because of carburetor icing in the humid, cold weather. When Sawina touched the pedal lightly, the ice on the throttle plate broke free but stuck in position. Sawina in trying to nudge it free might have pushed it to full throttle, where it stuck once more. He pressed repeatedly on the brake to stop the van, running down the vacuum reservoir in the brake system, and by the third push of the brakes he could not hold the van anymore. When and if this case comes to trial, the theory of the expert witness will have to explain the tire marks and what witnesses say they heard from the engine. These point to an immediate application of full throttle rather than several stages of throttle nudging and brake pushing.

The odds of pedal error go up when drivers are elderly, and also when drivers turn around in the seat to back their cars up, as in parallel parking. If sudden acceleration kicks in, drivers can always shift into neutral and turn off the ignition. The publicity about the brake light circuit has persuaded many police departments to drop the wigwag altogether. A better approach involves separately wired strobe lights that nestle in the brake-light lens, alongside the regular bulbs. This solution costs more but avoids all conflict with the brake-light circuitry. And besides, it's more attention-getting than the wigwag.

CONFUSION IN THE COCKPIT

According to the NHTSA report, the trigger at Holidazzle was missing the brake pedal and hitting the gas instead. Flipping the wrong switch or pressing the wrong button have been very common errors in the history of cockpits and control rooms. In the early days of World War II, when aircraft came out of factories in great numbers and even greater haste, the manufacturers found it expeditious to construct the electrical control panels out of identical, closely spaced switches. According to human-factors analyst Andre Chapanis, one common mistake caused by this design happened on the ground after a successful flight. While taxiing, B-17 bombers occasionally just flopped down on the concrete like a person might slump to the ground in a faint. Chapanis convinced the army air forces that pilots were being confused by identical toggle switches for landing gear and flaps, placed side by side in the cockpit. Pilots had been mistakenly raising the gear when intending to raise the flaps instead. His suggestion was to shape-code the switches, that is, to use a wheel-shaped knob for the landing gear switch and a flap-shaped knob for the other.

Human speech is rife with the possibility of error in dangerous places when people have only one chance to hear and comprehend an important message. Ordinary language is loaded with redundant information not because we love to repeat ourselves but because speech is often misunderstood. Repetition pounds the information through the barriers of noise and distraction.

Researchers Paul Fitts and R. E. Jones heard the following account while interviewing hundreds of fliers for their 1947 study of why so many aircraft were crashing and burning because of seemingly stupid mistakes in the cockpit. It happened at a flying school for military pilots. Two trainees went up in an AT-6 trainer, one serving as the safety man while the other worked on his instrument flying skills. Then they traded off, each pilot having a full set of controls at his own seat. At the end of the session the men left the training zone to fly back to base. The front pilot was steering, but then called to the rear pilot to take over. After a few minutes of level flight the airplane began to swoop to the left, entered a spiral, and then went into a vertical dive. The airplane pulled out below tree level, narrowly missing a trailer truck, but coming so close that the occupants stopped the vehicle and jumped out. The airplane climbed to two thousand feet, stalled, and dived again. Before it got close to the ground again, the front pilot angrily took the controls and flew the trainer back to base. As soon as the engine stopped, the front pilot challenged the rear pilot on his foolhardiness. "What," said the other, "I thought you were flying!" And so it happened that neither had been flying because the rear pilot had never heard the front pilot's request to take the controls.

Even written messages get messed up, as Miriam Safren and Alphonse Chapanis found at the Johns Hopkins Hospital during a 1950s study of systematic errors. At the time, doctors writing prescriptions by hand at the hospital often used Latin words or abbreviations for those words. One phrase for a prescription was *quaque nocte*, or "once every night," abbreviated as "q.n." on the prescription slips. Another was *quaque hora*, "once every hour," abbreviated as "q.h." on the slips. All it took for some patients to receive twenty-four times as much medicine in a day as recommended was for a pharmacist to see a handwritten "n" as an "h."

Prescription misunderstandings persist today, with the quantity "1.0" being mistaken for "10," or "OD" (once daily) being mistaken for the Latin abbreviation for right eye, *oculus dexter*. A 1999 report by the Institute of Medicine estimated hospital-staff mistakes as killing as many as one hundred thousand Americans each year. One of those medical mistakes happened in 1995 at a hospital in Stuart, Florida. A seven-year-old boy had been brought in for routine surgery but died in the hospital after two medications in unmarked cups got mixed up. A syringe intended for the local anesthetic lidocaine was filled with epinephrine instead, which caused the boy to have a heart attack.

When life and limb are at stake, it's proper to be excruciatingly clear when passing along important facts, even if that means repeating back instructions with military rigor. It may come across as an insult to assume another person is going to misunderstand, but the machine frontier is no place for worries about hurting tender feelings. When in doubt, check, because there are more ways for people to misunderstand than you can imagine. The book *Construction Failure* by Jacob Feld and Ken Carper describes what we might call the Case of the Missing Inch. In 1950 the drawings for a concrete retaining wall in Manhassett, New York, went to the builder with a line superimposed across the measurements specified for the vertical reinforcing bar, making a "1" vanish from view. Because no one caught the glitch, the wall went up with a quarter-inch-diameter bar instead of a one-and-a-quarter-inch steel bar. The wall fell in immediately upon backfilling.

AND THE GAS CAME SCREAMING OUT

The price of failing to communicate in a clear and timely way tolled on the night of July 6, 1988, when an explosion and fire aboard the *Piper Alpha* platform killed most of the crew. This giant steel platform, standing in 475-foot-deep waters in the North Sea, had two jobs before it melted and collapsed. The *Piper Alpha* drilled for oil like any rig—producing 140,000 barrels a day—and it also had a gas-processing plant aboard that purified huge volumes of natural gas piped to it from other platforms nearby. The *Piper Alpha* then pumped the finished goods to shore. On the afternoon of July 6, a day-shift crew doing routine maintenance work disconnected a safety valve from a pump that removed liquids from the natural gas, but they couldn't finish the job before the shift change. At the end of the shift, the men said later, they notified a supervisor that nobody should turn on the backup pump for the time being. The subsequent investigation indicated, however, that operators on the next shift who had control over this condensate system weren't aware of the warning. That evening the main pump went down and—surprise!—workers on the next shift tried to start the backup unit.

Gas flowed from an opening in the half-repaired pump, then exploded, knocking down firewalls. The uncontrolled flame began spreading to other parts of the oil and gas processing equipment. The jet of flame might have been a manageable problem had not the heat weakened the metal of the big vertical pipes bringing in gas from other platforms. These pipes were of about the diameter of tree trunks and carried two thousand pounds per square inch of pressure. When these "risers" let go, the flames shot hundreds of feet into the air and trapped people in the housing block at one end of the platform. The gas was "screaming like a banshee" as it shot out, one survivor said later. Another hour went by before the other plat-forms shut down their gas shipments to the *Piper Alpha*. Having no good escape route to the sea one hundred feet below, some workers stayed put and were killed by fumes and fire; others jumped. "It was a case of fry and die or jump and try," one survivor told a reporter.

In a scathing investigative report afterward, Lord Cullen went after the operator, Occidental Petroleum (U.K.), for its "superficial atti-tude" about safety precautions. No blast-resistant walls protected the worker housing, and the workers had no good escape route. The rig had a powerful set of automatic firefighting gear, but the system had been turned from automatic to manual to prevent divers from getting sucked into the intakes if the fire pumps came on. Finally, the owner had not followed up on warnings a year before, made during a safety audit, that clearly pointed out hazards from the gas risers and recom-mended steps to protect the pipes from breaking open in a fire.

Operator mistake on the scene was also part of the story. In a rather traditional sort of error, the maintenance crew failed to seal off the opening that their work had made in the gas pump. But equally important here was the new class of error, namely, the failure to com-municate clearly with fellow workers, especially during the handoff of work to another shift. It's an important skill, and easily underrated by those who think that working hard implies only the traditional type of activities, such as typing hard on a keyboard or running a power tool or driving a truck. The once-respectable attitude of "It's not going to happen on my watch" is clearly not justifiable anymore, not when so many tasks cross from one shift to another.

On the *Piper Alpha*, workers and managers had gotten sloppy about following the "permit-to-work" system, which was supposed to prevent the use of equipment that was in the midst of being repaired. Instead of holding a short but productive talk about high-risk repair work with the installation manager before getting under way, workers had gotten accustomed to throwing all permit-to-work papers on his desk. Covering all jobs big and little, this mound of paperwork swamped that key manager with unimportant checkoffs while mon-strous risks went sneaking by.

The garden-variety errors on the *Piper Alpha*—leaving a job half done and failing to make sure that the right people knew about an unsolved problem—would have been of little consequence had they happened two centuries ago. But in this time and this place, the consequences were extreme: 167 people dead, a $1.1 billion platform destroyed, and 12 percent of North Sea gas production cut off.

THAT LITTLE BLACK BOOK

Compare the risky habits that had developed on the *Piper Alpha* to how ordnance-disposal men communicated danger during World War II. As the war proceeded, these squads came across increasingly devilish timers and booby traps, which the enemy added to warheads in the hopes of wiping out the small cadre of experts. These men worked slowly, inspected often, and were fastidious about calling out each step they took to assistants positioned out of danger. If they survived the job, they prepared meticulous field notebooks that the armed forces immediately published for other squads in the form of urgent bulletins. Each man knew the information would help others, regardless of whether he himself survived the job.

The library at the Aberdeen Testing Center in Maryland has a collection of these bulletins. Typical is an April 1944 warning about the German "Tellermine," an antitank land mine. The bulletin warned men to tie a 150-foot rope to each mine and drag it off for blowing up with demolition charges. Some Tellermines had a booby trap set to detonate the explosive if anyone lifted it more than an inch, and some were set on a hair trigger. Said the bulletin, "Do not attempt to disarm it, because shear pin igniter may be partially sheared and in a very dangerous condition."

In this same tradition of wisdom keeping, operators at safety-minded petrochemical plants maintain a thorough log of trouble-shooting reports and unresolved issues, sometimes called the "black book." Boeing scrupulously preserves its safety and design lessons from decades past, gathering them into a highly confidential book titled *Design Objectives and Criteria*. This book is consulted during the early stages of planning for each new airliner model, and it's here that a designer would learn how Boeing solved many hundreds of flight problems through the years—problems that are still relevant.

With this kind of knowledge to back them up, people are more likely to remember the consequences of ordinary error in extraordinary places. We can reduce the incidence of these mistakes but not eliminate them. As with the DC-10's cargo doors, equipment is no good unless its designers have planned to handle a wide range of errors in the field.

ERRORS EVERYWHERE

Usually we can get through our days making many errors but never having to pay the bill. Studies of airliner crews have shown that they make dozens of small errors on a long flight, like mishearing a controller's command until it is repeated, putting a sophisticated fly-by-wire airliner in the wrong mode, or climbing past the desired altitude when flying manually. Even Captain Bryce McCormick on that miracle day in the DC-10 over Windsor had made mistakes. He'd erroneously concluded that his control over the trim stabilizer on the tail wasn't working. It had indeed been working, but the response to emergency controls was so slow he hadn't realized it. And after landing, the copilot had realized before McCormick had that a change in the thrust reverser setting would steer the airplane away from the fire station it had been heading for.

A good system, and operators with good "crew resource management" skills, can tolerate mistakes and malfunctions amazingly well. Some call it luck, but it's really a matter of resilience and redundancy.

We know from cockpit records that surprisingly small problems can make for fatal distractions if this resiliency factor is absent. Early proof of its importance came on December 29, 1972, during the flight of Eastern Airlines Flight 401. The crew of this L-1011 jumbo jet had clear weather on the nighttime approach to Miami, and everything looked good until a green lamp on the landing gear status panel failed to light up after the command to lower and lock the gear. Approach Control approved the request to maintain at two thousand feet and to set the plane on autopilot while the captain, copilot, and engineer checked out the problem on the nose gear. The copilot pulled out an electronics assembly holding the bulb to check and replace the light.

The problem was indeed a burned-out bulb, but the crew didn't know that. They also didn't have a good visual cue about their altitude because they were heading west from Miami, over the Everglades. In checking the bulb, one of them bumped a control column and caused the autopilot to click off. Distraction ruled, with the copilot telling the others that the light-bulb assembly had jammed when he tried to push it back in. Since they couldn't check the new bulb, the captain ordered the engineer to go below to the compartment under the cockpit to check the nose gear mechanism by eye. But the engineer came back to report he couldn't see the mechanism in the darkness. More problems with the electronics; once again the engineer went down to check the nose gear. During the three minutes that the captain and copilot spent discussing the jammed light assembly, a musical warning chime went off, alerting anyone who was listening that the aircraft was more than 250 feet below its assigned altitude. Apparently, no one heard it.

The last chance to do something came when the approach controller noticed the airplane dropping out of its assigned altitude of two thousand feet. On the one hand the controller had assigned an altitude and the pilot was leaving it without permission, and so this discrepancy was something he was entitled to point out; but on the other hand the pilot in command has the sole responsibility of flying the airplane. The controller compromised his way out of the dilemma by radioing Flight 401 and asking, "How are things comin' along out there?" in the hope that this would prompt the crew to check the situation before calling back. At this point there were four people in Flight 401's cockpit—the three flight crew members and an Eastern maintenance specialist who had been riding in the jump seat—and all were messing with the light assembly.

The controller's radio call only prompted the reply from 401 that the crew wanted to turn back to Miami. A half minute after the controller's call the airplane crashed into the swamp. Seven seconds earlier the copilot had told the others that the airplane was too low, but there was no time left to do anything about it.

Good training in situational awareness can overcome the ancient survival habits that narrow our attention in a crisis. It shows up in many places, but it all adds up to remembering where you are and what you are supposed to be doing. One of the most impressive examples of good training is that of air traffic controllers. They maintain a sort of mental map—called "the picture"—of the direction and altitude of all the airplanes under their control. They are so good at this task that they can continue to function even if their radar scopes go down.

READY FOR WAR

While I was visiting the missile-alert command center at the Cheyenne Mountain Complex in Colorado, the duty officer pointed out a display of various three-digit numbers; he called it the "real world status board." As he explained it, the board can display in shorthand form many dozens of possible situations that might occur in the outside world and that have some bearing on the work at the complex. Each listed situation has its own three-digit code, starting from broad categories (those digits are hundreds) and going to finer detail (tens and ones), which is akin to the way in which the Dewey decimal system is used to organize books. A space shuttle entering orbit has one code; an impending shuttle return from orbit has another code; a Russian launch yet another.

Even good situational awareness is not enough, though, if a subordinate notices something important but hesitates to raise the subject with the one in charge. A case in point is that of the air traffic controller who radioed Flight 401 to ask how things were going rather than stating his specific concern, namely, that the airplane was heading toward the ground without clearance. A simple statement by the controller that the airplane was dropping out of its assigned altitude, if sent in time, might have made all the difference.

To retired airline pilot Frank Tullo, chair of the Air Transport Association's Human Factors Committee, a close study of thirty-seven recent airline crashes reveals a striking pattern. Tullo says the following factors show up entirely too often: the captain was flying the aircraft at the time of impact; the flight was behind schedule; and the flight had an inexperienced copilot who was not pushy about a safety problem. "The safest thing in a crew is an assertive, skilled copilot," Tullo says.

Many of us are trained from childhood to get along, to defer to authority, to make friends instead of waves, to say nothing if we can't say something nice. It was this force at work on that British Midlands flight when passengers saw the left-hand engine belching flame and smoke, but said nothing after the captain told them over the intercom that the right-hand engine was the one having problems.

When the path of least resistance is to step back and let things run their course, it takes a special type of person to insist that he knows more than the authorities about the problem at hand. Studies show that Australian copilots are better at assertiveness than those of many other nationalities, but they aren't born that way, which means anybody can learn to speak up. If schools and Scout programs can teach children to resist abuse, surely we can teach adults to assert themselves with safety in mind.

Not that it will be easy. In a simple but thorough set of psychological investigations at Yale in 1961, Stanley Milgram discovered that most of the people he tested were entirely willing to give somebody else multiple strong electric shocks if the commands to continue came from an authority figure, in this case a white-coated "scientist" who told them that shocks were necessary as part of a memory and learning experiment. The shocks were not real, and the forty-seven-year-old "victim" was following a script, but for all Milgram's subjects knew, the jolts were bringing a man to unconsciousness and perhaps even death. Milgram found immediately that subjects were so willing to continue straight through the series of thirty shocks, rising from 15 to 450 volts, that he had to change the experiment to have the victim yell and plead for release as the shocks went up. Nevertheless, some participants continued to administer the shock all the way to the end of the series even after the "victim" went completely silent, apparently unconscious or dead.

Many of Milgram's subjects were unhappy and tense about what they were doing: pausing, fretting, questioning, but then going ahead under pressure of orders. One of the comparative few who stopped and flatly refused to continue, despite repeated demands from the "scientist," was a thirty-one-year-old German-born woman who had spent part of her adolescence under Hitler's rule.

Not a single subject simply stood up and walked out. One way that some people made themselves feel better was to try and excel with the technical details, concentrating on which switch to push and for how long. "This is, perhaps, the most fundamental lesson of our study," Milgram wrote, "[that] ordinary people, simply doing their jobs, and without any particular hostility on their part, can become agents in a terrible destructive process. . . . A substantial proportion of people do what they are told to do, irrespective of the content of the act and without limitations of conscience, so long as they perceive that the command comes from a legitimate authority."

People have a right and an obligation to stand up and protect their own lives and the lives of others, especially if they are responsible for those lives. Perhaps it would help if we had a social convention similar to the navy's "permission to speak freely, sir!"

TIME TO THINK

So if we have people with both good situational awareness and crew communications, and who have been trained to stand up and be counted, is anything lacking? Such people can manage a complex and confusing emergency only if they have time to think before having to act, and a few minutes might be all they need (although ten minutes would be much better!).

Part of the trick in high-fear situations is knowing what needs to be done immediately, what can wait, and which actions cannot be reversed after second thoughts. A memorable example of the latter came in January

1874, at the bottom of the seventeen-hundred-foot-deep main shaft of the Ophir Mine of Virginia City, Nevada. Four men had the job of deepening the shaft by drilling holes and packing them with dynamite. On this particular job, the blasters were using slow-burning fuses to set off the charge. The miners loaded the explosives into four holes, packing the top of each hole with sand and gravel, such that only the fuses peeked out. They lit the ends, leaving themselves a couple of minutes to get to safety. One man pulled sharply on the rope that would tell the operator at the surface to run his steam engine so that the man-cage, which would carry the men to safety, would be dropped to the bottom. Nothing happened after several increasingly energetic yanks on the rope. The rope had snagged on a timber somewhere above. The men looked at each other and then at the fuses. Three men stood paralyzed, but the fourth began scrabbling at the dirt to pull the fuses loose. He managed to pull two of the fuses, but the flame on two others vanished below the surface, burning down toward the charges. He yelled at the others to start climbing as fast as they could. Three men got away with minor injuries; the fourth got a rock planted in his skull but survived.

Recall what Bryce McCormick did when things looked very bad: explosive decompression, engine fire alarms, plane going into a turning dive, rudder jammed hard over, jet engine thrust down to idle, and the horizontal stabilizer handle broken off in his hand. He leveled his airplane and then he paused to think things through with his co-pilot and engineer. Together they rebuilt their world and then they started again.

Our missile attack response plans are founded on this gimme-time-to-think premise. In the event of an attack by an intercontinental ballistic missile, the Pentagon believes that emergency action controllers, war-room commanders, and eventually the president should be able to squeeze out enough time—about four to six minutes—to verify that a threat is real before deciding to launch a counterattack.

Perhaps . . . at least until other countries have hypersonic missiles. Being pressed too hard by time is worrisome because when harried we draw only on human memory, and that's a weak reed in times of stress. Quick memory, the short-term kind we use for phone numbers, holds less than ten random items. Moving facts from short-term into long-term memory requires extra mental effort. Human long-term memory has enormous capacity, but reliability is a serious problem: memories fade or, worse, get jumbled together with false recollections so that what we think we know for sure is subtly but dangerously altered.

Some manual skills stay with us for life, such as the way our muscles and brain learn to keep us upright while riding a bike, but the high-level thinking skills required to operate and troubleshoot a vast machine need frequent refreshing, with updates to incorporate new lessons learned. The navy spends millions to keep its submarine crews sharp on such "perishable" skills at its training school at the submarine base in Groton, Connecticut.

Because of the long-recognized shortcomings of memory, operators of any complex system are supposed to follow elaborate checklists when starting or shutting down or while taking the machinery through any important transition. We know from technological disasters that transitions in their broadest sense—from aircraft landings to factory crew changes—are the times of greatest danger. But errors can happen with operators using a well-written checklist if they skip over steps or lose their place because of distractions.

Even routine operations offer rich opportunities for error, given the short attention span of most people. It's hard to sit and watch a machine all day to make sure it stays within limits. After a half hour or so of reasonably attentive behavior, the mind drifts to ponder something more interesting. Later, if a crisis develops, the operator has to spend valuable minutes trying to figure out what the machine was doing at the point it went haywire. Sometimes it's impossible, because by the time the human realizes something is wrong, there's not a minute to lose.

For all their excellent features, autopilot systems have led to such incidents. According to former NTSB Air Safety chief C. O. Miller, an Aerospatiale ATR-42 transport airplane plummeted to the ground in October 1987 while flying through the Italian Alps because the autopilot had been adjusting the flight controls to keep the airplane in the air despite thickening ice on the ailerons. The pilots didn't detect the growing problem until the automated flight controls had exhausted their ability to keep the wings level. The airplane began to roll off on one side, and this shut down the autopilot. Nothing the pilots could do could save the situation at this point, so the airplane crashed. But had a human been operating the controls instead of the autopilot as the ice built up, he would have known of the impending crisis well before it peaked, from the deteriorating "feel" in how the aircraft handled.

A modern cause for what at first appears to be machine madness is that the computer-driven system is in a different "mode" than the operator thinks it is. This was the case with Eastern Flight 401 before it crashed into the Everglades; the pilots thought the autopilot was keeping altitude, but the autopilot had given altitude control back to the pilots after someone inadvertently nudged the yoke. Several crashes with the Airbus A320 have seen pilots fighting the controls, with humans trying to land their aircraft when the machines wanted to climb the aircraft away from the airport.

But it's also true that airliners don't crash very often, perhaps because people in a cockpit appreciate the precariousness of their position. For a long time, it was much easier for people working at chemical plants to forget the consequences of simple human error. Trevor Kletz, in studying mistakes and disasters in petrochemical plants, reminds the managers he talks to that their initial actions after the report of trouble say volumes to workers. This becomes part of a company's "common law" and is much more persuasive than safety manuals or mission statements. After a fire or explosion, Kletz asks, is the first response from the top a flat denial of all blame followed by an order to get production going, or is it a budget sufficient to fix all problems before production resumes? Dana-Farber Cancer Institute in Boston is so serious about encouraging workers to report errors that managers have sent thank-you notes to those who come forward.

BOREDOM BUSTERS

Even when incentives and a safety-minded workplace culture are in place, some skepticism is still in order. Not everyone will respond as

desired as managers "turn the dials" in their organizations. One reason
is the last class of human error to be addressed in this chapter, something that psychologist Frank Farley of Temple University calls the
"Type T" personality. These are thrill seekers, or people who can't live
without uncertainty. Consider the working style of Billy Hamilton.
He's long dead, but his style lives on.

In 1855, Hamilton came aboard the newly built Mississippi River
steamboat *Fanny Harris* as assistant engineer. In overseeing the engines,
Hamilton alternated with the chief engineer so that one of them was
on duty at all times. Hamilton quickly won a reputation on board for
his skill in running the engines, his edgy practical jokes, and his eager-
ness to join the first mate in a fistfight with drunken or rebellious
deckhands. Hamilton kept a small iron cannon on the boat, firing it on
holidays or sometimes just for fun in the middle of the night. Adding to
the excitement, Mississippi steamboats often raced each other up and
down the river. Most races were impromptu, starting when two vessels
happened to be going in the same direction at the same place on the
river.

Billy Hamilton hated to lose a race, and apparently hated to go slow
at all. If the captain wasn't paying attention, Hamilton would build a ver-
itable wood-fired blast furnace under the boilers, turning up the forced-
air blowers until solid chunks of live coals blew high out of the twin
chimneys. And it took more than that to bring the steam pressure up:
Hamilton rigged pulleys so he could make quick changes to the way in
which a fifty-pound anvil balanced the pressure safety valve. It was a fla-
grant violation of the federal steamboat law.

As the captain got to know Hamilton's style, he began sending
his steward down occasionally to peek at the pressure gauge when
Hamilton was on watch. Hamilton responded by building a sheet-lead
cover for the gauge. It left only a short length of the indicator needle
exposed, with no markings visible. In that way Hamilton would know
the boiler pressure but the captain wouldn't. Of course the day came
when Hamilton saw the steward trying to pull the lead cover off and
get a good reading. Hamilton had a two-pound shaping hammer handy
and threw it at the man's head, knocking him unconscious. The steward

recovered, the captain gave up, and the lead cover came off. And this, with Hamilton knowing full well that if the boiler blew, he would be the first to go.

Risk takers come across as bigger than life, so we tend to remember them fondly as long as they don't kill too many people along the way. Take John Luther Jones, a locomotive engineer for the Illinois Central Railroad. You may have heard of him by the nickname "Casey" Jones, after his hometown of Cayce, Kentucky. Early on the morning of April 30, 1900, Jones was highballing engine number 638 and six passenger cars at speeds of up to one hundred miles per hour, trying to make up a delay in his departure from Memphis. Near Vaughan, Mississippi, Jones heard warnings and saw the tail end of a train ahead, sticking out from a siding. The fireman jumped, but Jones stayed to hold the brake. The train slowed, but not enough. The locomotive plowed through the cars and Jones died in the wreck.

Still Jones lives on in the "Ballad of Casey Jones." This folk song celebrates his decision to stay in the cab and says nary an ill word about the reckless speeds that morning. Or about the railroad bosses who so valued schedules that they'd reward those people who took almost any chance to keep the trains running on time. Operating decisions like this—the unwritten law of the company—can turn a reasonably safe system into something very dangerous, one day at a time.

11: ROBBING THE PILLAR

SLACKING OFF WITH THE HIGH-POWER SYSTEM

As dawn broke on December 3, 1984, a person in a plane flying low over the northern half of the industrial city of Bhopal, India, could have thought that the filming of a new Michael Crichton terror-virus movie was under way. Thousands of people lay along the streets, on sidewalks, and around transportation hubs. One of the victims was the station master of the Bhopal railway station. A few hours before, despite suffering from poison gas, he remained at the station to send radio messages to incoming trains, warning them to stop short of the city. His efforts kept thousands of passengers at a safe distance from an obscure and money-losing herbicide plant that had just become the site of the worst chemical disaster of all time.

Although the geographic origin of the deadly cloud of methyl isocyanate was immediately clear, the cause of the Bhopal disaster was not. There are at least three theories as to what triggered the chemical release.

There is the Indian government's official explanation, as well as two other stories.

An image to explain what happened at Bhopal came to me while touring an underground coal mine called the Pioneer Tunnel, preserved as a historical site for tourists at Ashland, Pennsylvania. During the mine tour, the guide talked about an old practice called "robbing the pillar." In regions having horizontal layers of coal, miners would blast and muck out a room but leave massive pillars of coal at regular intervals to shore up the roof. When most of the coal was out of a room, a few men went back to shave the pillars to get as much coal out as they could. To survive, a miner needed to stay in close touch with anyone else working in the room, and each man needed to know how much the other guy was weakening the pillars across the room at the same time. Working too independently, together they would bring down the roof.

The safety of high-power technology stands on pillars, too, and they can be weakened at any time during the lifetime of a system, from design through construction, from routine operations through disposal. Here we look at the maintenance side, the uniquely human jobs of wrench wielding and troubleshooting that can never fall to computerization in the way that other jobs have. Maintenance is the soft underbelly of the system, the open door to disaster.

It's human nature to let things slide. Delayed maintenance is part of our trial-and-error, stretch-it-out style. We also look to cases where people shaved away at the margin of safety by shutting off safety devices or letting them break without replacement. Typically, each person assumed that enough slack remained to avoid causing a safety problem. And so it goes, people "working together" to crash a system.

WHERE'S WARREN?

According to the Indian government, one of the people personally responsible for the worst case of pillar robbing in history is Warren M. Anderson. India still wants him for culpable homicide and has notified Interpol that he is a fugitive from justice. Back in December 1984,

Anderson was the chairman of Union Carbide Corporation. That's the month when the gas leak from a plant owned by its subsidiary killed about seven thousand residents of Bhopal. Like everything about the disaster, these numbers are disputed. But it is safe to say that the number of those injured by methyl isocyanate and other toxic gases has probably surpassed two hundred thousand. Whatever its exact toll, the disaster was every manufacturer's nightmare: something toxic got across the fence and attacked the neighbors.

Though repeatedly warned by advisers and lawyers not to go to India, Anderson went anyway, arriving at the Bhopal airport three days after the accident. But his display of empathy and apology did nothing to mitigate the public anger as authorities hustled him from the airport and placed him under house arrest. He departed after six hours when the Indian subsidiary paid his bail. Later, Union Carbide also paid $470 million in settlements after being sued in civil court by the Indian government. As part of the deal, Union Carbide sold off its ownership interests in the subsidiary.

At last report, Anderson remains at large, in ill health and living in secluded retirement at Vero Beach, Florida. He refuses to stand trial in India. The Bhopal district court is willing to try Anderson in absentia, but it has no way of punishing him, short of kidnapping him and flying him to India. Meanwhile scribblings on Bhopal walls, and speeches in continuing public demonstrations, call for his execution.

To other Union Carbide executives, India's hounding of Warren Anderson is just political posturing and harassment. They say that the Indian courts originally agreed to drop criminal charges as part of the $470 million civil settlement. Union Carbide paid the money within ten days of the agreement, but the court nevertheless revived the charges. And it wasn't their fault, anyway, according to Union Carbide: sabotage was the cause of the runaway reaction.

By the month of the disaster, December 1984, this parent was a diversified multinational. Union Carbide's consumer division sold Glad trash bags, Prestone antifreeze, and Eveready batteries. The chemicals division was the world's seventh biggest supplier of industrial chemicals. Operations in forty countries contributed almost a third of total sales.

Union Carbide (India) Limited (UCIL) was one of those sub-
sidiaries, and 1984 was its fiftieth anniversary year. Managing UCIL
through the company's eastern international division in Hong Kong,
Union Carbide owned a shade more than half of UCIL's stock. Thousands
of Indian shareholders held the rest, with government-run insurance
companies owning most of that. Working out of thirteen Indian factories,
UCIL was mainly a battery maker but also manufactured herbicides, elec-
trodes, plastics, and raw chemicals.

One of those factories was in Bhopal, in the central forested state
of Mahdya Pradesh. It had opened in 1969 as a herbicide mixing and
packaging facility two miles north of the central railroad station. The
banner products were Sevin and Temik, herbicides that agricultural
experts regarded as much safer for the environment than the wildlife-
killing DDT. They were part of the "green revolution" of chemical
tools and hybrid plants that would feed the world's hungry. By 1978,
UCIL was making Sevin using imported ingredients. One of the raw
materials was a toxic liquid called methyl isocyanate, or MIC (pro-
nounced "mick"). It arrived in stainless steel drums from Union Carbide's
plant in Institute, West Virginia. MIC was an ingredient for a chemical
called carbaryl, the active weed-killing component for its products.

Facing a crowded market for herbicides and needing to cut costs,
UCIL erected a processing plant at Bhopal in 1979 to make its own
MIC and other key ingredients. This unit was in production by 1980.
The plant stored the MIC in two big tanks as a liquid, with nitrogen
pressurizing the empty space left at the top of the tank. A third tank
was for storing MIC that didn't measure up and needed reprocessing.

UCIL reduced the MIC production crew by half from 1980 to
1984. Apparently they did so at a price: a corporate safety survey in
May 1982 had warned of slipshod maintenance practices at Bhopal,
such as workers leaving work unfinished. UCIL reported back later to
the Danbury headquarters that it had solved nearly every problem
listed in the safety report.

The city of Bhopal didn't like the idea of MIC being made at the
plant, given the additional hazards that manufacturing it posed, but the
state and national government endorsed the new operation. Mean-

while, the destitute residents of Bhopal had been erecting thousands of ramshackle dwellings all the way up to the street running along the south side of the UCIL plant.

The Bhopal example illustrates how easy it is to end up with dense housing next to a plant that manufactures deadly chemicals. A company sets up a plant to make a product that is fairly innocuous. But after a while the owner needs to expand production to include the manufacture of a hazardous chemical. To do so safely requires that the factory have an appreciable perimeter of uninhabited space in case of toxic release or explosion. But by now people live all around the factory and they won't move, and the local economy has far too much at stake for "what if" arguments to stop the project. The developing world has had many such shantytown disasters; two of the bigger ones happened in the same year as Bhopal. One was in Brazil and the other was in Mexico. Both were fires caused by flammable liquids and gases, and together they killed a thousand people.

At Bhopal—when things were working right—fresh MIC came out of a distillery vessel and went through stainless-steel pipes into storage tanks. A dirt and concrete cap covered all three tanks, which were equipped with sensors to collect information about volume, temperature, and pressure. When the plant needed MIC, nitrogen gas under pressure pushed the liquid out of the storage tanks and through pipes to a production reactor. Any MIC that failed quality tests because of contamination went over to the third storage tank, or workers could direct it straight to a "vent gas scrubber tower" for chemical destruction.

When the plant first opened in 1980, the vent gas scrubber tower was one of four safety systems that could help manage a MIC leak to some degree. The scrubber tower was a hollow cylinder about six feet wide and sixty feet high. Water and caustic soda sprinkled down from the top like an indoor waterfall. This water would attack any MIC gas coming up from the bottom and break it into harmless compounds.

The tanks also had a refrigeration system that kept the MIC at a low enough temperature to discourage a runaway reaction in which enough heat and pressure would build up to force mass quantities of poison to gush through the safety valve. Bhopal's MIC unit also had a big torch, to burn off escaping MIC vapor. The fourth safety system consisted of a water spray, which was intended mainly for fire fighting but thought to be useful in the event of an MIC release (because water would neutralize some of the vapor if it escaped).

In June 1984, the barrier to disaster formed by these safety systems began to fall. The first to go was the cooling system for the tanks, because workers siphoned off the refrigerant for use elsewhere on the grounds. Though not as famous as other problems leading to the disaster, according to some analysts the refrigeration shutdown may have been the worst mistake in the whole chain. This is because MIC that's warm is much more likely to react than MIC kept near the freezing point of water. Had the MIC been kept cool, contaminated water leaking into the tank from elsewhere in the complex plumbing system would have been much less likely to trigger a runaway reaction.

The Bhopal production reactor made its last MIC in October 1984, leaving tank E610 with forty-two tons in storage and tank E611 with twenty tons. At this point workers turned off another line of defense, the scrubber tower. Later they shut down the flare as well, to replace a corroded section of pipe. They would have to wait for the new piece to arrive from the United States.

Shutting off the safety devices made sense if a person believed that the only danger would be during the MIC production run, rather than while the finished product sat around in storage. Shutting down the safeties saved the company some cash, and this was not a minor consideration for a plant that was running at only 40 percent capacity and losing lots of money. (But as we know it cost them much more in the long run.)

The tale of mechanical woe continued in November, with both main MIC tanks having trouble maintaining their nitrogen gas pressure. Nitrogen was not only necessary to pump the liquid out of the tanks, it was the inert gas that protected the liquid MIC from chemical contamination. Without nitrogen gas keeping these chemicals at bay, they could start to seep into the plumbing. One contaminant was alkaline water, with traces of metals, leaking from the scrubber tower. It trickled through the pipes and reacted with MIC vapors to form a plastic gunk that gathered on the pipe walls.

At the end of November the workers decided to repair the nitrogen systems on just one of the tanks, E611, because it had enough MIC to meet their needs for now. They would repair tank E610 at a later date, but as it turned out, E610 would be the tank feeding the disaster.

According to the Indian government, this is what happened late on December 2, 1984. (Union Carbide has its saga of sabotage to tell and we will hear that afterward.) The task that night was for production workers to clean out the gunky trimer that had been accumulating in pipes near the tanks. It was important to keep water from getting into the MIC tanks in this operation, and because valves leak or can be set incorrectly, the textbook procedure was to insert metal barriers in the pipes (called "blinds") to block off any stray water from leaking into the storage tanks. Then the operators would run water into the pipes and wash out the trimer. Because of a foul-up and an unfilled maintenance supervisor position, no one installed the metal barriers, which would have protected tank E610 from all possibility of wash water leaking through the pipes. Once the cleaning started at 9:30 P.M., drains that were supposed to let water out of the pipe clogged so badly that the water level rose in the pipes instead of draining out the bottom. One of the workers pointed out that wash water from the pipe-cleaning effort should have been coming out the drains but wasn't; nevertheless, a supervisor brought over from a UCIL battery factory told him to keep going. Water rose high enough to reach a pipe to a pressure-relief valve, twenty feet off the ground. Then, seeking any kind of outlet, the water worked its way downhill near tank E610 to another valve that was supposed to be closed but wasn't. Nobody knows why it was open, but it could have been a mistake made by someone trying to work on the tank's leaky-nitrogen problem. Leaks and human errors with valves happen regularly in the chemical industry; that's why the metal barrier was supposed to be in place.

All these mishaps worked together, the Indian investigation concluded, to let more than a hundred gallons of wash water into tank

E610, sometime after 10:00 P.M. A heat-producing chemical reaction of water and MIC began in the tank about this time.

After the shift changed at 10:45 P.M., incoming operator Suman Dey noted the pressure in the E610 tank: about ten pounds per square inch. That was higher than what it was supposed to be, about two pounds, but still within limits. Near midnight, workers at the plant knew MIC was leaking from somewhere because it had an effect like tear-gas fumes, and they were feeling it. One man reported seeing water and MIC coming out of a pipe. After the routine tea break, Dey knew the tank pressure was off the scale. He left the control room to see what was happening. He got close enough to hear rumblings from the tank and the sound of gas and liquid escaping from a valve. Back in the control room he tried to bring the vent gas scrubber off standby, but according to the instruments it wouldn't respond.

The flare that could have burned the MIC vapor was out of service, and operators couldn't cool the tank because the refrigerant was gone. That left only the water sprays. The plant's firefighters discovered they couldn't shoot water high enough to make any difference because the gas was coming out from a stack one hundred feet off the ground, and the water didn't reach that high.

Venting for two hours, MIC hugged the ground and spread southward, toward the railroad station. MIC attacks the mucous linings of the respiratory system, causing much the same effects as the chlorine gas that Germany first used on Algerian troops in April 1915. Nobody outside the plant had gas masks, of course, but had they been warned beforehand about MIC, people might have known to lay down in the lowest place they could find, covering their faces with wet cloth. This crude measure could have saved many because the water would have neutralized much of the effect before MIC reached their mucous membranes.

Instead, terrified by police loudspeakers that told people to flee from poison gas, two hundred thousand people ran for the hills, or tried to. Many died in their houses or along the way, choking on MIC and other toxic gases spewed off in the runaway reaction.

OTHER SUSPECTS

Union Carbide and its consultant Arthur D. Little Inc. dispute just about all of this account except the fact that tons of MIC did escape, causing death and injury. According to company spokesman Thomas Sprick, "There is no truth to various allegations that the tragedy was caused by design defects, insufficient personnel and neglect by plant management." The company says that a MIC-process-unit trainee was so angry over being demoted that he deliberately piped water into tank E610, setting off an exothermic reaction that overpressurized the container and pushed much of the stored chemical out a vent pipe. Union Carbide and consultant say that the combination of pipes and valves hypothesized by the Indian government could not have put wash water into the tank and was not supported by the status of the piping system when examined afterward. They say only deliberate action could have caused the disaster.

The Union Carbide argument makes some sense in strictly mechanical terms, but the human angle challenges the imagination. And only imagination can fill the gaps, since the company has never named the perpetrator—though the company managers have said they know their man's identity. "For Union Carbide to identify the employee now would serve no useful purpose," Sprick told me. Did the man intend to hit the company in the pocketbook by ruining a batch of MIC? There were simpler and safer ways to cause more cash damage. Or did he intend to kill and injure thousands? He would have had to coordinate his lunatic scheme with the company's shutdown of multiple MIC safety systems at the plant, starting six months before.

There's a third possibility, some researchers like Paul Shrivastava say, that fits the mystery rather well: there was a problem in tank E610 as the shift began, predating the washing operation. It had come from a slow leak of contaminated water into the tank, possibly aggravated by problems with the nitrogen system. The tank was headed for a runaway reaction that night; first signaled by rumbling and venting from the tank. This in turn caused a worker to panic, and he connected a hose to run water into the tank, in a totally misguided attempt to "cool off" the contents.

People can argue through the centuries about what first triggered a catastrophe, but it's much harder to hide from evidence of lax maintenance and crumbling safeguards. At Bhopal the plant operators had cut three of four safety systems that could have stopped or at least lessened the destruction. The official story from Union Carbide at this time is that maintenance standards at Bhopal had nothing to do with the disaster, but some officials have tried to put distance between headquarters and the behavior of the Indian subsidiary. As Union Carbide vice president Jackson B. Browning told a writer for the *Atlantic* afterward, "There were maintenance problems that would not have been tolerated at a plant in the United States."

DISPOSABLE STEAMBOATS

Trying to keep pressure in check is a concern as old as the machine frontier. Back in the steam age, engineers knew that any of several acts of carelessness could burst a boiler. Letting the water level drop too low was as dangerous as running pressure too high because without enough water in the boiler to absorb the incoming heat, the furnace fires would soften the metal until pressure tore apart the weakened boiler wall. And poor repairs on a boiler wall could destroy a boat if the patch was weaker than the engineers assumed. Letting mud and scale build up in the boiler was also detrimental because it blocked the easy transfer of heat through the metal to the water. Finally, dumping cold water in a boiler too fast could crack the metal and trigger an explosion.

It was just such a careless act that blew up the steamboat *Sultana*, which was carrying an ungodly large crowd of released Union prisoners at the time. As documented in *The* Sultana *Tragedy* by Jerry Potter, it was not the *Titanic* tragedy of its age—it was worse, probably killing well over eighteen hundred people.

Mississippi steamboats, with an average life of five years, were close to expendable. Some engines passed through two and even three steamboats, salvaged from hulks and then repaired. Names were recycled, too, as with the line of steamboats named *Sultana*. The word *sultana* refers to a sultan's wife, or alternatively his sister or mother. After the first boat with the name, the others of the *Sultana* line demonstrated that the common riverboat's life was short and brutish. The second *Sultana* collided with the steamboat *Marie* in November 1846. The third *Sultana* burned to the water's edge at a St. Louis wharf in June 1851. The fourth burned in March 1857.

We are interested in the fifth *Sultana*, launched from Litherbury's Boat Yard of Cincinnati. It was the biggest of the line at 260 feet long. It was strikingly narrow, at forty-two feet of beam; people of the era thought a narrow boat would save fuel and gain speed. Even with a thousand tons of cargo it drew less than three feet of water. Based on her size, federal law allowed the *Sultana* to carry 376 passengers plus the crew, which amounted to about 460 people.

Federal law also set pressure limits on her high-pressure steam power plant, with the intention of making her safer than ones built before the law was passed in 1852. That law, the U.S. Steamboat Code, was passed by Congress despite strong industry opposition because of a rash of disasters so bad that it aroused the concern of the American public, which at the time was remarkably tolerant of danger. During the years between 1816 and 1848, 233 explosions on American steamboats had killed more than two thousand people. As English visitor Philip Hone said, steam had apparently arrived just in time to take the place of war.

Four riveted, iron boilers supplied the *Sultana*'s engines with steam that the law capped at 145 pounds per square inch based on the metal and its thickness. Two big pistons, each two feet in diameter, drew steam from the boilers. The slow pushing and pulling of each piston drove a paddle wheel at about twenty revolutions per minute.

The boilers lay side by side, lengthwise with the boat, each eighteen feet long and almost four feet in diameter. There was a safety valve and even a failsafe. Part of the piping to the safety gauge was supposed to lock open and begin dumping steam at 150 psi. These were "tubular boilers," meaning that many tubes ran through the inside, making a path for hot gas from the furnace to travel through the boiler and give up more heat. The smoke started at the firebox in the front, moved to the back end along the full length of the boilers, and then came forward again by traveling along the tubes inside the boiler. Fire heated such a boiler from both outside and inside. Then the smoke went up the chimneys.

Pipes connected the boilers together. The boilers lay so close to each other, running parallel and lengthwise with the boat, that a problem affecting one boiler was bound to affect the others. Because of the greater surface area of a tubular boiler compared to the old fashioned ones, people in the steamboat business knew it was more likely to explode. To be safe, engineers needed to watch the water level closely and to clean out minerals and mud according to a strict schedule. Cleaning was difficult with the kind of boilers the *Sultana* had, because the tubes ran in a zigzag pattern.

On May 9, 1863, the *Sultana* raced two other steamboats from Memphis, Tennessee, to Cairo, Illinois. The one named *Belle Memphis* pulled away and beat the others easily. The captain of the *Belle Memphis*, J. Cass Mason, had gotten into trouble with federal authorities earlier, when a gunboat crew found his steamer *Rowena* carrying quinine and three thousand pairs of uniform pants for the Confederacy. As a penalty, the Union commandeered the *Rowena* for as long as they would need it in the war effort.

In March 1864, Mason and partners bought the *Sultana* and hired a captain to run it, but Mason took over soon after as master of the boat, probably to keep costs down. Mason's financial problems got worse. He had started as the majority owner of the *Sultana*, but by the spring of 1865 he had sold off everything except a one-sixteenth share.

Mason joined the *Sultana* with other steamboat owners in a business consortium that won an open-ended contract to haul Union soldiers for five dollars a head. Union officers brought twice that amount. By April 12, 1865, the *Sultana* was ready to leave St. Louis for New Orleans to continue hauling Union men under the contract, along with any other freight and passengers Mason could attract. But he had to wait one more day for the federal steamboat inspectors to finish. The two inspectors ran the boiler pressure up to 210 pounds per square inch, checked out other essentials, then signed off on the safety certificate. On April 13, the *Sultana* started its run down to New Orleans. On the way south it stopped at Vicksburg, because Captain Mason wanted to make sure he would have a northbound cargo when he returned in two weeks.

His intended cargo waited impatiently four miles inland from Vicksburg, at a holding area called Camp Fisk. Here were thousands of Union prisoners that had survived the bits of hell on earth called Camp Sumter near Andersonville, Georgia, and Castle Morgan at Cahaba, Mississippi. The prison at Andersonville had opened on February 24, 1864. At its peak, it held thirty-three thousand men, three times its intended capacity. Prisoner Joseph Stevens counted 150 men dying every day. The same wagons that carried out the dead carried in the food, mostly spoiled pork and ground-up corn and corncobs.

"One might have thought that the grave and sea had given up their dead," wrote one prisoner, recalling the sight as fifty-five hundred men emerged from the gates of Andersonville in March 1865, all tattered clothes, skin, and bones. When the Union men arrived at Camp Fisk, they were only partially freed, needing official confirmation of the parole agreement between Confederate and Union armies before they could leave for points north.

While stopped in Vicksburg, Captain Mason called on two important Union officers. One was a former riverboat captain named Morgan Smith. Smith, now a brigadier general, was commander of the Vicksburg post, and he promised to put aside a full load of prisoners for the *Sultana's* return trip. The other was a man named Reuben Hatch. Hatch had once used his brother's political connections to Abraham Lincoln to escape a court-martial after being accused of taking bribes as an assistant quartermaster for the Department of the Mississippi. A military commission had later tested Hatch's skills for and knowledge of the position, reporting him "totally unfit" for the job he had already held for four years. However, at the time he met Mason he was *chief* quartermaster for the Department of the Mississippi, and he promised the captain plenty of prisoners.

The *Sultana* continued downriver to New Orleans, laid over two days, and headed back north on April 21, with passengers and cargo. Less than a day from Vicksburg, a small crack opened in the center-left boiler, and high-pressure steam jetted out. Worse, the iron in the area was bulging. The *Sultana* had had boiler leaks on the two previous trips, and the situation was now getting worse. Engineer Nathan Wintringer cut the speed down and continued to the Vicksburg landing, expecting to undertake thorough repairs there.

By the time the *Sultana* arrived, two thousand prisoners had already shipped out from Vicksburg on two other riverboats. Mason fumed. Those were supposed to be the *Sultana's* prisoners. He demanded a full cargo of prisoners immediately but was told that it would take days to prepare the official roster for the next boatload. As Mason stormed about the town, he stopped to straighten out the boilermaker that his engineer had rounded up to repair the cracked boiler.

The boilermaker's name was R. G. Taylor. Taylor, twenty-eight years in the trade, was of the opinion that at the minimum he needed to pull out two iron plates on the cracked boiler, a significant operation taking days. Each boiler was made up of riveted iron plates that overlapped at the joints. He would need to chisel out the rivets connecting two of the plates, cut and bend new plates of the right size, and rivet them in. Nothing less would do, Taylor said. In fact, he later added that all four of the *Sultana's* boilers showed overheating damage. This damage could have come from either failing to clean the boilers often enough or from letting the water supply run too low.

Mason talked Taylor out of replacing the plates. Taylor agreed to put an iron patch over the bulging cracked spot now if Mason promised to bring the steamboat in for repairs in St. Louis later. The iron patch Taylor used on the job was thinner than the boiler walls on the *Sultana*, by about one-ninth of an inch. It took the better part of a day to complete the job. In the rush, one thing was overlooked: no one reset the safety valve to a lower pressure (100 pounds per square inch, instead of 145 pounds) to compensate for the thinner and therefore weaker boiler wall where the patch was. According to Gene E. Salecker, author of *Disaster on the Mississippi*, this would prove a critical omission.

Now things were speeding up for the *Sultana's* trip north. Army captain Frederic Speed agreed that Mason's boat could have all the remaining prisoners at Camp Fisk, which the Vicksburg post estimated to be fourteen hundred men. And the Union officers would not require the boat to wait for an official roster. In good can-do fashion, the guard detail could put that list together while the boat was under way. After two other steamboats arrived, one of them larger than the *Sultana* and both able to take many hundreds of prisoners, Captain William Kerns protested strongly about putting them all on the *Sultana*. Kerns also went to Hatch and Smith, asking them to divide up the load. But Hatch and Smith held fast to the "all-on-*Sultana*" plan. The other two boats left, taking no prisoners.

That left the *Sultana* with plenty. First the boat collected hundreds of prisoners from the hospital, then three trainloads from Camp Fisk. The first estimate of fourteen hundred prisoners—which would have put the boat four times over the legal capacity—was grossly underestimated. At least twenty-one hundred former prisoners, and possibly more than twenty-two hundred, crammed on board. There would have been even more passengers but for the intervention of a Union doctor, who steered away another 278 men from the hospital and had two dozen men on cots carried off the *Sultana* before it left.

As the passenger count approached two thousand, even Captain Mason thought the boat's capacity had now exceeded its safety limit, but having started the overboarding process he could not stop it. The army had a long history of overloading steamboats over the captain's objections. Including the crew members and other passengers, some of whom belonged to the Chicago Opera Troupe, the *Sultana* had approximately twenty-three hundred people on board and perhaps twenty-five hundred by Potter's estimate. Many of them stood or sat on the upper decks, which would make the boat top-heavy. The crew jammed in lumber to brace the floors against collapse, and Mason warned the passengers against rushing to one side of the boat.

The *Sultana* left Vicksburg at 9:00 P.M. on April 24, heading upriver. If it made the trip, the chief clerk said to a passenger, it would be "the greatest trip ever on the western waters, as there were more people on board than ever were carried on one boat on the Mississippi River." In addition to the passengers, the boat carried horses, mules, and pigs and a cargo of sugar and wine. There was also a seven-foot-long alligator on board—in a crate under a stairway—which served as a mascot.

For the next two days the engineers fretted about the patch on the center-left boiler, checking it often. The opera troupe got off the boat at Memphis and the *Sultana* continued on toward Cairo, Illinois. Chief engineer Wintringer left the engine room at midnight, telling the assistant engineer on the graveyard shift that the machinery was running well. The steam pressure was, allegedly, 135 pounds per square inch. At 2:00 A.M. on April 27, seven miles upriver from Memphis, three of the *Sultana*'s boilers, including the one that had been patched at Vicksburg, exploded as the boat approached Island No. 40.

The burst of steam and iron fragments tore a hole through the cabins behind the engine room, caving in the decks and trapping some survivors. The light timber of the boat caught fire quickly, and within twenty minutes the boat was engulfed. The two chimneys stayed upright and then fell opposite ways, one crushing what was left of the pilothouse. Though an exact count is impossible, at least fifteen hundred civilians and soldiers died in the explosion or by drowning, but the total was probably more than eighteen hundred. Even the boat's pet alligator died, stabbed by a soldier who wanted its crate to use as a raft. According to the army's inquiry, the explosion was caused by a shortage of water in the boilers, which could have been aggravated by the tilting of the top-heavy boat as it made a turn. This would have drained water from the high-side boilers. The engineer in charge at the time, who survived the blast, insisted that he had checked the water level and it was adequate. The director of the federal Steamboat Inspection Service pinned the blame on the patch. Because its metal was thinner than the boiler walls, the safety valve should have been set to open forty-five pounds per square inch lower than usual to compensate for the difference.

There aren't many mea culpas uttered in the world of industrial catastrophes. No matter what a government agency or court has to say, companies in such a fix will beg to differ. Just as Union Carbide blamed the Bhopal disaster on a disgruntled employee, one of the owners of the *Sultana* blamed its catastrophe on Rebel saboteurs.

It takes constant support from top management to keep an organization running safe, says Richard Jones, vice president of industrial insurer Hartford Steam Boiler. "If the attitude comes only from the bottom up, it will be snuffed out."

12: MACHINE MAN

SURVIVING AND THRIVING ON THE NEW FRONTIER

In 1886 the tramp steamer *Trojan* arrived at a small port north of Liverpool to take on cargo. The captain, Jonathan Barber, had left for Cornwall to spend time with his family, and so first mate Pryce Mitchell was in charge. Because this was an iron-making area, the *Trojan's* cargo would consist of hundreds of tons of massive steel bars, consigned for Mobile, Alabama. Each bar, called a "bloom," was eight feet long, a foot wide, and six inches thick, enough to roll into a single railroad rail. The accepted way to load blooms aboard a small ship like the *Trojan* was awkward and dangerous. The men stood the blooms on end in the hold, tying them together and packing them three layers tall into the cargo hatches so that the cargo formed squarish columns of solid steel, twenty-five feet high, rising to the top of each hatch like square pegs in square holes.

The crew's next job was to spike and wedge enough wooden braces together to make sure the steel didn't come loose at sea. Securing cargo is the unsung part of maritime work, but among the most important. Many ships over the years have been sunk after their cargo ran wild, and the crew considered improperly stacked steel about as dangerous as it gets. If the blooms worked loose as the ship heaved and pitched, the havoc they would cause in crashing and sliding around would send the *Trojan* down in minutes. It would be worse than a loose cannon from the old man-of-war days, when a wheeled gun after breaking loose from its chocks and ropes could smash holes in the ship's sides if not trapped right away.

First Mate Mitchell put in a request to the company's marine superintendent, a man named Tranter, for plenty of strong timbers. Tranter answered that he already had plenty of wood lined up. Mitchell was horrified by what arrived dockside: a pile of old planks, castoffs from a ship in the process of being refitted for a new deck. Mitchell challenged Tranter about this cost-saving measure and prompted the reply that so many worried underlings have heard: if Mitchell didn't like the plan, he could find a new job. There were plenty more men who needed work. It looked a perfect dilemma: Mitchell paid the living expenses for his family and so felt he couldn't quit. Yet Mitchell also knew the planks could not hold this cargo together, but he had no way of buying enough timbers himself.

Mitchell wrote to Captain Barber, who replied that Mitchell should go along with the superintendent until he got back. The night Barber arrived, he inspected the braces built of planks and agreed that the steel would come loose at the first gale and sink the ship. With the ship scheduled to leave on the morning tide, Barber pulled the boatswain aside and told him to find the harbor authorities as soon as possible and ask them to inspect the ship, without revealing that the plea came from the officers. In the morning the boatswain went off in search of help as the crew made ready for sea. With Tranter standing on the pier waving good-bye but no authorities yet in sight, Mitchell told the engineer to find enough problems to stall for half an hour. Finally the authorities arrived and ordered the cheap bracing ripped out and replaced with sound timbers. The *Trojan* arrived safely at Mobile, despite storms so bad that of the five ships that left the port bound for Mobile, over a ten-day period, with cargo loads of steel, only two made it. The other three ships sunk somewhere along the way.

Given the many warning memos sent and then ignored prior to other disasters, it appears that in some cases a guerrilla tactic like the one on board the *Trojan* is going to be the only way to stop a dangerously flawed machine from going into service. If you find that possibility deeply troubling, ask yourself whose lives will be at stake if something goes terribly wrong.

MACHINE STORM

What that Atlantic storm did to the steel-hauling ships has another message in it, too. When I began this project some friends asked what anybody could boil out of the huge variety of technological disasters we've seen. I didn't know. Late in the process one thing occurred to me, triggered by the story of the *Trojan*: some of the worst accidents in the field had a flawed technology meeting an unexpectedly strong force of nature. Here are the four ingredients for a really bad day of this kind: A lot of people let some machine take them into a place where their lives depend entirely on its proper workings. This hunk of technology has problems so serious they start showing up even under benign conditions. The people in charge don't follow up on problem reports from line staff. Finally, some natural force like an earthquake or storm arrives to tear away the facade of safety.

It also sounds a lot like the wreck of the *Ocean Ranger*. And like the typhoon-induced river flooding in August 1975 that burst the mighty Banqiao and Shimantan Dams in China, killing an estimated twenty-six thousand people in a single night. And like the collapses of the bridge on the River Tay and the snow-packed roof over the Hartford Civic Center.

Given the number of near misses, our machines hold up well most of the time. Perhaps one distant day evolutionists will look back to our time and say that this was when *homo sapiens* began evolving into *homo machina*, "machine man," a species able to understand what it really takes to build and run complex, high-power systems in a world with forces that are still a lot more powerful than we are.

If so, they might trace one of the early appearances of the new species to Admiral Hyman G. Rickover. According to political reporter Ron Brownstein of the *Los Angeles Times*, while Americans like to hear stories about brilliant, idiosyncratic leaders who can spur people to excellence, society can't run this way. Brownstein was referring to the occasional principal or teacher who can turn a failing school around in record time, but his observation applies equally to mastering technology. Rickover was such an idiosyncratic genius if there ever was one, and although his career spanned the record length of sixty-three years, even he had to retire and die.

LEADER OF MEN, LEADER OF MACHINES

Instead of a Rickover replacement we really need entire banks of workers and managers who feel personally responsible for a good outcome. In early 1968 astronaut Wally Schirra dropped in to see how final work was going on the command module of his *Apollo 7* spacecraft. Though doing his best to avoid bumping into or stepping on anything in the capsule, Schirra's knee pressed on a bundle of wires. A woman who was working in the spacecraft at the time reached over and slapped him smartly, saying, "Don't you dare touch those wires. Don't you know we lost three men?" After finding out that it was one of the astronauts she had struck, she said she was terribly embarrassed. Schirra replied, "Don't be. I want people like you working on this spacecraft."

Some people do better at building this attitude in the ranks than others. Lord Thomson was a bold and inspiring leader, but the same qualities that made him that way—unshakable self-confidence and the ability to motivate people into accomplishing things they otherwise wouldn't have tried—made for trouble when he used those skills to rush the *R.101* into service, against strong advice from the people closest to the project. He was indeed a leader of men, but that didn't make him a leader of machines.

John Novomesky, of the IBM roof-fixing incident in Brazil, says that his West Point years taught a management lesson that explains why Les Robertson went down again and again to Brazil, to check on the repair work. "You have to know where the decisive point is," Novomesky says. "That's where a leader is supposed to be."

It's as true in mastering technology as in running a battle. One of the heroes of the 1939 rescue of three dozen men out of the sunken submarine *Squalus* was Admiral Cyrus Cole, commandant of the Portsmouth Naval Yard. Cole had been planning to take visitors on a tour of the *Squalus* upon its return to base from a test dive that day; immediately after hearing of the emergency, he hurried to the docks and found another submarine, the *Sculpin*. Cole personally directed its commander to proceed to the last known whereabouts of the *Squalus*. He summoned "Swede" Momsen's rescue unit. And after the *Sculpin* saw a distress signal and sent back word, Cole got aboard a tug and went out to supervise the situation personally.

Contrast this with the events leading up to the August 1907 fatal collapse of the south arm of the Quebec Bridge. The consulting engineer was Theodore Cooper. In his younger years he had done marvelous things. As an inspector and then a construction supervisor on the history-making St. Louis "Eads" Bridge in the 1870s, Cooper had done his profession proud by spotting serious problems early, once staying up for almost three days to make sure things got done right. But he had passed his prime when he took on the Quebec Bridge job.

Cooper made three trips to the Quebec Bridge site beforehand, but by 1904, when steel erection got under way for the vast span over the St. Lawrence River, Cooper was too old and unwilling to travel. Cooper never visited the project again, relying on mail and telegrams to keep him in touch with the construction of his design, a design that had been forced to the edges of safety by financial difficulties in the Quebec Bridge Company. When signs of trouble began appearing—steel ribs not lining up, because of the higher-than-estimated weight of the structure—Cooper wasn't there to see it firsthand, as he had been with the Eads Bridge. When the steel began buckling and the alignment problems grew worse by the day, Cooper heard about it but could do no more than send worried telegrams. His last telegram tried to order the crews to add no more weight to the bridge until after investigation, but he'd sent it via the steel fabricator's factory, and the message didn't make it to the construction site in time. Five hours after he sent the message, on the afternoon of August 29, 1907, seventy-five workers died as nineteen thousand tons of steel came crashing down.

Sometimes it takes some persuasion to get leaders to take their places at a critical time and location. In the early days of airmail flying, the mail pilots came to believe that their crash rate was unacceptable, even for people accustomed to danger. Finally a group of them convinced the U.S. Air Mail Service that postal supervisors at the airports were ordering them aloft in bad storms and poor visibility. The solution? Not a new regulation spelling out what weather was safe and unsafe, but rather this simple order: if an outgoing pilot desired, his supervisor had to join him in the cockpit to fly a circuit around the airport before the pilot went off on his mail run. Quickly the supervisors' tolerance for bad weather dropped. Admiral Rickover made it a rule that he or his top assistants would be riding on every nuclear submarine during its initial sea trials, so they'd have as much at stake as the men who had to be there. It focused attention; it kept people sharp.

CLOSE CALLS

We've seen some extremely close calls over the years, where people acted to stop the chain of disaster with minutes, or even seconds, to spare. We writers love last-minute escapes, but it would be a serious mistake to depend on the ability of operators and pilots to snatch survival out of the jaws of a disaster that somebody else started. The breakdowns as they hit their peaks may be too chaotic and weird to understand, much less control, in a just a few minutes. In their last unhappy hours of life in the sunken *Thetis*, machinists and engineers labored heroically without proper tools to jury-rig pipes and pumps to lighten their boat enough to bring the stern to the surface again, but in the end their efforts saved no one.

The most important time period is long before a fire or explosion, at the design and planning stage. It's why "hell-fighter" companies like Boots and Coots have shifted their focus away from fighting oil-well fires and blowouts toward selling expertise to oil companies so these mishaps don't happen in the first place.

People planning new or upgraded chemical factories should know the importance of what safety consultant Trevor Kletz calls inherently safe design. At Bhopal, an inherently safer plant wouldn't have had a problem with methyl isocyanate storage tanks because the plant wouldn't have had any such tanks at all. One way is to run such chemicals straight through, without storing them as intermediates for later use. Or better yet, a safety-conscious designer would have used one of the other carbaryl-producing processes that didn't need MIC.

Consider another early decision, whether to standardize equipment. Consistency, say some, is the sign of a small mind, but it's a real virtue with high-power technology. Having lots of the same devices out in the field lets us compile a record of experience. Further, standardizing controls and machinery allows operators to shift easily from one plant to another, or pilots to move safely from one cockpit to another. Among the reasons why Southwest Airlines of Dallas, Texas, has built such a remarkable safety record (no fatal crashes, ever, in thirty years of short-haul, time-critical flights) was its decision to standardize on the Boeing 737 jet. Pilots in standard airplanes can more readily spot anything out of the ordinary and are less likely to make a "control-confusion" mistake in the cockpit. Mechanics are less likely to use the wrong part or tool.

Such early planning to avoid cockpit crises will become even more important in the future, because complexity will increase. That means it's going to be harder for people on the scene to fix anything in just a few minutes. Will that time be available as systems run closer to the edge to raise efficiency for their owners? We know that even the most canny and prepared people need time for their judgment skills to work. They need time to look for discordant information and they need to check with others who might have a crucial insight. In an ideal world, a system would be able to hold itself together for at least five to ten minutes during an emergency so people on the scene would have time for a sound decision. Sometimes they'd know that the better part of valor is to stand back and let a computer handle the job.

LIKE A RUNNING BACK

In researching the habits of high-reliability organizations, Karlene Roberts and Karl Weick decided that good operators plan for a way out. As one example, just before touching down on a carrier deck, a navy pilot knows to shove his thrust levers full forward, so he will have power to get off the deck if his tailhook misses all the arresting wires.

An experienced outdoorsman in Nevada's Black Rock Desert once told me that he engages his four-wheel drive only sparingly; if he kept it on all the time he would eventually meet a quicksand mire that he could not get out of because he had already burned up all his safety reserve. "I use the four-wheel to get out of trouble, not into it," he said.

Again and again, workers in high-hazard jobs have told me how they never go into a chancy situation without planning beforehand how to get out, such as firefighters who do not begin confined-space rescues until they have breathing packs to wear and buddies to pull them out if something goes wrong.

A drill ship I visited had escape hatches all over the vessel, each painted red. All rooms over a certain size had to have two exits. It's the principle that air traffic controller Tony Brescia once told me about his job: "You always have an alternate plan. Like a good running back, when one hole is closed, you look for the next." Escape routes are part of preserving a little safety margin for the really bad day.

Is that kind of advance planning proof that high-reliability organizations have really arrived? Or would it make any difference for the complex and interactive system that Charles Perrow worries about? For myself, I make no arguments that anything can be made perfectly safe, nor do I know what "safe enough" means. In a national emergency the American public eagerly puts up with risks they'd never tolerate otherwise. Put me down as one who thinks we'll have to settle for systems that have problems on a regular basis. If the systems are resilient and the workers have support from the top, problems will likely stop well short of disaster—most of the time. But where the consequences are irreversible and final, such as an accidental nuclear war, like Scott Sagan I find it hard to believe that we'll be able to keep our collective finger on this hair trigger indefinitely without twitching even once.

Even the best-run systems always have something off-line or running out of tolerance, out there in the wilderness of high-pressure piping, wires, and cable trays. No force on earth can get everything to stay in balance all the time. To insist on perfection is to shut the whole thing off. And the people who run the systems wouldn't pay attention, anyway. As sailors say, this would be seen as another stupid order from "the beach," meaning from people who don't know how the machine works out in the theater of action and haven't the courage or will to master it.

THE SEVEN RULES OF RICKOVER

When called in to testify on the organizational lessons from the Three Mile Island reactor meltdown, Admiral Rickover described seven principles of safe operations for reactors. The first was a rising standard of quality as time went on, and well beyond the minimum required for licensing or permitting. Second, the people running the system should be highly capable, trained by those who have run the equipment under a wide range of conditions. Third, supervisors in the field have to face bad news when it comes, and that means taking problems up to a level high enough to mobilize enough effort and talent to fix those problems. Fourth, Rickover said, people on the job needed to have a healthy respect for the dangers of radiation. Fifth, training must be constant and rigorous. Sixth, all the functions of repair, quality control, safety, and technical support must fit together. One way to do that is to have highly placed people dropping in on the plant, particularly on late shifts, during maintenance downtime or when the plant is shifting from one state to another. And seventh, the organization must have the ability and willingness to learn from mistakes of the past.

Owners and managers of high-power systems not being run along these lines should beware of the trouble ahead. In a disaster key employees will be lost, business will stop, reputations will suffer, and any shortcuts that managers ordered along the way are going to be seen in a very bad light, possibly to the point of criminal indictments.

The costs of a single bad day can be enormous. Phillips Petroleum incurred total costs of $1.4 billion in damages and business interruption plus government penalties from a fuel-air explosion, on October 23, 1989, that killed twenty-three workers at its polyethylene plastic factory in Pasadena, Texas. It happened because contract workers were doing maintenance work under a plastic-manufacturing pressure vessel and misconnected a hose to the compressed-air system controlling a valve. The ball valve opened, dumping tons of flammable hydrocarbon vapor into the air. When the gas found a flame, the explosion had the force of two tons of TNT.

I suggest that companies think of their time on the machine frontier as a privilege, bestowed by the rest of us. It's a valuable location because working the frontier offers more opportunity for profit and growth than old technological territory. In our era, some frontiers with seemingly great potential include genetic engineering, artificial intelligence, and deepwater exploration for oil and gas. But holding a place on the machine frontier is not a constitutional right. Losing the privilege could happen if a laboratory lets something loose onto the citizenry or if some cost-cutting decision causes the destruction of a valued habitat. One mishap might not cause public support to collapse, but two incidents could be more than enough.

It's naive for companies to think that the public is going to forget a disaster a few months later. Memories fade, but so does the public's willingness to put up with recklessness. Some communities are now canny enough to insist on legally binding "good neighbor agreements" before a potentially hazardous plant locates there or before an existing plant gets a permit to expand. Such agreements, like one signed for a Unocal refinery in Rodeo, California, requires the company to pay for precautionary items such as an independent expert to audit the plant for chemical safety, an independent health-risk assessment, and vans to transport the injured in case of a dangerous chemical release. Without written agreements, citizens have found in cases like Exxon's Prince William Sound oil spill, promises will not be kept.

SHUTTING THE DOOR

The public anger aroused by catastrophes can shut down programs for years, and in some cases, for good. After the crash of the airship *R.101* in 1930, the British scrapped their dirigible development forever. An eight-hundred-square-mile slick from an offshore oil well blowout off Santa Barbara, California, in 1969 is the reason that California renewed a ban on new drilling leases in 1990. In time, a near disaster at the Enrico Fermi-1 liquid metal fast breeder reactor on October 5, 1966, put an end to plans for building more commercial breeder reactors all around this country.

Even if it were possible to reach perfection, it wouldn't matter for more than a short time, anyway. All complex systems mutate under the pressure of technology and business, so what was good last week is different now. Every launch of the space shuttle is a little different from the last one because there are hundreds of suppliers in the chain, and they all have to change materials and techniques now and then. It's probably better not to believe in our own perfection, anyway, because it's such hubris that leads to the belief that failure is impossible.

Indeed, as a matter of mathematics any multiple-failure chain of circumstance is probably going to look exceedingly improbable before it happens. But viewed another way, these chain-of-failure events were not impossible in the sense that their occurrence flouted any laws of physics. They did require a long line of coincidences all coming together. So how could such a statistically remote event happen? In Charles Perrow's terms, as systems get more complex and tightly coupled, there are more combinations that can lead to failure. Think of this as the Powerball lottery's evil twin. In a multistate lottery the odds of any given person winning are extremely remote, but the likelihood that *someone* is going win, sooner or later, is certain. Many of the mishaps we prefer to regard as impossible aren't impossible at all—they just take longer.

"PAINFULLY CLEAR" IS GOOD

Survival on the machine frontier requires that we speak and write with extreme clarity, even if doing so necessitates asking people to repeat back what you told them, or taking a tone of voice that could come across as bossy. We also know that most people will try to avoid making trouble, particularly any trouble visible to outsiders, even though they are convinced that a catastrophe is near. Line workers might make a few comments to a foreman but drop the subject. Underlings send strong memos to overlings. Rarely do we see people going outside the organization to enlist help, out of a conviction that they have a personal responsibility to act.

A few reasons come to mind about why a seriously worried worker is likely to get back in line after making a show of concern. Maybe he's wrong, and the machine will do fine after all. There's the risk of being fired for insubordination. Even without this fear, the average employee is going to have some second thoughts about causing trouble for the people who hired him or for his coworkers. A sharply worded internal memo, then, is a feel-good solution. It puts the worker on record as one of the good guys without challenging the structure of power and obedience that Stanley Milgram's experiments revealed.

We know from a long list of catastrophes that warning memos from inside didn't make any difference. That would include the *R.101*, the *Challenger*, Three Mile Island, Bhopal, the DC-10 and its cargo doors, *Apollo 1*, and others. The mistake made by people who stop at memos is that they think their writing will stand on its own—that its hard truth will somehow vanquish opposition as quickly as Dorothy Gale's bucket of water melted the Wicked Witch. But a warning memo is more likely to annoy the boss than correct any wrongdoing or negligence. Opponents to action will mobilize, and their best arguments won't be countered by a piece of paper.

To supervisors and managers, such a memo offers a lose-lose situation. If they endorse the message, it's going to cause trouble "upstairs," but if they go on record as skeptical it could cause big trouble for them if a disaster does occur. The bureaucratic solution is to let the memo sit in the inbox for a while—then send it back for more explanation.

For employees willing to take a position that could jeopardize their jobs, there are ways around this barrier of indifference. It's scary, but it is this risk that brings credibility to the argument. Whether or not whistleblower-protection statutes apply, managers are going to take very seriously a clear warning that arrives from a respected, seasoned employee who stands ready to quit his job if nothing is done. Of course, that's presuming the employee has also done his homework and goes beyond a mere memo. That means he can clearly explain the consequences of failure, he can answer every challenge from highly motivated opponents of equal expertise, he has gathered his allies early, and he has requested face-to-face meetings with managers who have authority to make critical decisions.

If you've been in a motor vehicle accident, you've probably relived that crash for weeks or months afterward, thinking of many things you could have done differently to prevent it. In my own case, this obsessing continued for several weeks and was part of the grieving process. Being in the line of responsibility for a system failure is sort of like that, except the memory never wears off. There were so many chances to do things differently, so many warnings shrugged off.

You might find that gambling your livelihood is one of the best chances you've ever taken. You might even find that your fears of being fired for insubordination were exaggerated. William LeMessurier blew the whistle on his own project to make sure repairs were made; he ended up enhancing his stature. In the 1950s test pilot Bob Hoover refused orders from North American Aviation management to continue testing the supersonic F-100 fighter, saying that he would not fly it again until the company rebuilt the tail, because it was unstable at high speeds. In the end, North American and the air force modified all the aircraft, as Hoover requested, and he stayed on.

One person can make a difference if the official process requires that person's signature to proceed and if he is willing to risk his job by refusing to sign off on the required documents. And, most painfully, if he's willing to make trouble for the people who hired him and/or whom he calls his friends. None of us can say beforehand with any certainty whether we'd have the guts to do so.

Consider this: if you were in the direct line of authority when things went terribly wrong, some of that same unleashed destruction could end up in your office, anyway. Recall that after the *Apollo 1* fire, mighty Joe Shea got the boot out of the Apollo program. An official court of inquiry following the Tay Bridge disaster in 1879 laid most of the blame on engineer-builder Sir Thomas Bouch, sending him to an early grave and turning his name into a British synonym for messing things up.

Stopping a project juggernaut is very stressful, and surely we'd rather the buck pass on, but be aware that more of these tough calls are in your future. Like it or not, troubleshooting and the exercise of judgment are among the few uniquely human skills in running high-power, narrow-margin technology.

LOOKING FOR THAT GOLDEN KEY

In the late 1950s, a Livermore Laboratory researcher named John Foster envisioned a device that would, without fail, distinguish between wartime and peacetime. Mounted on warheads, this "golden key" would stop inadvertent detonation of a nuclear weapon. Foster never got beyond the prototype stage because the military feared it would malfunction, and in so doing interfere with their control over the nation's arsenal.

The idea of a device with unerring judgment about what's really going on, one immune to all stress and distraction, is as appealing as ever. The air force and navy are working on a computerized "Pilot's Associate" to sort through cockpit information to tell pilots in single-seat combat planes just what they need to know—a sort of computer-aided situation-awareness tool. I can see such a device working on airplanes and to the benefit of air traffic control and other complex systems, but on a new project under construction, will computers have the ability to look at the whole picture, the gestalt, and see what is missing or wrong? Probably not. Indeed, the time has long passed when any single person could do so. Only a group of people, communicating well, can manage it. According to Michael Collins, one thing that put the *Apollo 1* spacecraft at risk was poor discipline concerning changes. The wiring system had been altered and expanded so many times, without documentation, that no one had caught up with all the changes by the time of the fire.

A dock outside the Submarine Force Museum at Groton, Connecticut, is the final resting place for the *Nautilus*, the first nuclear-powered submarine. After the self-guided tour and on their way out, visitors pass through the enlisted men's mess room. A small but interesting placard is posted on a table leg: THE TEN COMMANDMENTS OF DAMAGE CONTROL. It offers good advice, such as "Keep your ship watertight," "Have confidence in the ship's ability to withstand severe damage," and "Practice personal damage control." The ninth commandment is "Take every possible step to save the ship as long as a bit of hope remains," and the last one urges sailors to "keep cool; don't give up the ship!"

NOT THE DAY TO DIE

These are all ways to increase the zone of survivability. Sometimes expanding the zone is just a matter of attitude and conditioning rather than one of more safety hardware. The men of the *Thetis* were prisoners not just of a pressure hull but also of the belief that they could not survive an ascent from 160 feet without breathing gear, thinking that they would die from the bends. In fact, we know now that they would have done better to clear their sinuses, brace themselves for some excruciating pain, and cycle out of the submarine from the escape hatch, heading for the surface with only the air in their lungs, exhaling on the way up. Years after the disaster, the Royal Navy discovered from some surprising experiments on H.M.S. *Osiris* that volunteers had been able to leave a submarine's escape trunk as deep as 600 feet and float up to the surface without wearing any pressurized breathing gear. All the air they needed to rise to the surface could be held in an "escape hood" attached over the upper body. While there was some risk of injury—ruptured eardrums were most likely—the odds of survival were good, particularly at shallower depths. The U.S. Navy now prepares its submariners to escape using the "free ascent" technique if necessary.

My point is that after machines have left us stranded and at the edge of destruction, it's time to put the pain aside and attempt desperate measures. We know from many survival accounts that sheer stamina and a driving desire to cling to life count for a great deal. Do you think that death comes in minutes from near-freezing water? It's less likely to happen to those who refuse to give up. A cook from the *Titanic* survived in that frigid water for two hours, until rescue; a twenty-eight-year-old seaman who fell off a freighter on Lake Superior in 1999 swam to shore for six hours, through forty-six-degree water and eight-foot waves. Scott Richards's secret was swimming without pause, which kept his metabolism working.

Though collected figures are hard to come by, it appears from news accounts that dozens of people have survived falls of more than two hundred feet without an open parachute. A flight attendant fell thirty-three thousand feet from a DC-9 over Czechoslovakia in 1972 and survived. On May 19, 1991, skydiver Jill Shields landed in a swamp after her two chutes failed to open, ending up with a fractured pelvis and three broken vertebrae. Though the odds are clearly bad, there's a chance of survival if a person knows to use a spread-eagled posture to hold down his speed and steers himself toward a flat impact on vegetation or on soft ground, like a plowed field. Posture makes the difference between a possibly survivable flat impact at a hundred miles per hour or a certainly fatal one headfirst at three hundred miles per hour. As the tenth commandment on the *Nautilus* says, don't give up while life remains.

I'VE GOT A SECRET

Unfortunately, it seems many companies and military organizations and government agencies would agree on one more commandment, the

Eleventh Commandment of Damage Control: Conceal all problems that might bring on trouble for the organization. It's easy to rationalize: it was a freak occurrence, but we immediately fixed any problems; people should leave us alone about it because nothing like this will ever happen again, to us or anyone else. A full two years after the half-billion-dollar explosion of Kansas City Power & Light's Hawthorn Plant, the Missouri Public Service Commission filed an accident report but it had no worthwhile safety insights to impart because the company had insisted on confidentiality for all relevant information. Twenty-two years after TMI-2's meltdown, there is no definitive word on what went wrong with the pilot operated relief valve because lawyers are still struggling over who is liable.

As part of a court settlement in 1981, the John Hancock life insurance firm made great efforts to seal records and block its hired experts from ever discussing what went wrong with their skyscraper in Boston, made notorious by the media in the 1970s because of plate-glass windows that kept popping out. The record sealing was unfortunate because the building had problems more serious than window popping, and open discussion of the structural problems and solutions would have helped engineers around the world.

Certain things are not widely discussed outside the engineering community but probably should be. One is the effect of salt corrosion on concrete parking ramps, which can cause one deck to collapse and bring all the decks below it down as well. More chilling is what structural engineers call the "P-delta moment." It refers to how the overturning tendency of a skyscraper can be magnified by its sway under certain wind conditions. All tall buildings sway in a wind, but some are unusually "limber" and sway more than others, and this tendency throws the building's weight off center. In theory the P-delta problem (which is an entirely different matter from the Citicorp scare) can tip over a building if the unlucky combination of wind speed and wind direction comes together. One engineer who did not want to be identified said Hurricane Alicia in 1983 set up a P-delta condition that brought one Houston skyscraper "close to instability." The building's flexibility convinced owners to stiffen up the building's frame despite considerable cost.

Sometimes the habit of hiding embarrassing news goes to desperate extremes. After an explosion in a sixteen-inch gun turret aboard the battleship *Iowa*, which killed forty-seven men, a navy inquiry first blamed a sailor's sabotage. The navy had to retract this claim after Sandia National Laboratory reported that problems with the propellant and improper ramming procedures appeared to be the cause. According to the *Los Angeles Times*, for eight years the public didn't hear about the deaths attributed to Firestone tires blowing out on Ford Explorers because courts were sealing the trial records. I'm convinced that if certain chemical factories could be put on ships out in the middle of the ocean, some companies would do so just to stay out of the public eye.

If we were all spies working inside hostile territory, we'd be well advised to protect important information by sealing ourselves off in "cell" organizations. When people organize themselves into spy cells, none of the members except those at the top know much. In the most extreme case, each person is his own cell and does not know the real names of anyone else in the organization. A cell organization guards against capture and interrogation, because while no person can resist expert interrogation for long, he can't reveal what he doesn't know. But sealing off vital information is a bad practice on the machine frontier. As one step, we could start looking for insight from the many hundreds of narrowly averted near disasters.

THE WAY OF THE MACHINE

Some groups have recognized the truth-telling power of the near miss. The Aviation Safety Reporting System is simple and effective: Workers in the aviation industry are required to file anonymous reports about every incident they know of—whether minor or major—that could pose a threat to safety. It might be a mistake by pilots or air traffic controllers, a mechanical malfunction, or a false reading on instruments. To encourage people in the industry to come forward, FAA enforcement cannot use the information from the tips called in. NASA has been operating the reporting system since 1976. During a typical year,

NASA receives thirty-five thousand anonymous reports, which retired pilots and air controllers study to identify trends. Information gets to the front lines in the publication entitled *Callback* and as safety bulletins. The Veterans Administration has hired NASA to set up a similar system for detecting medical errors in its hospitals.

Opening up such firsthand, hard-won information has been a long step toward achieving what we might call a state of enlightenment, something like the satori of Zen Buddhism. Enlightened people are more likely to ask the tough, informed questions that any good organization needs and should welcome.

Armed with such information, and a healthy touch of fear, operators and pilots and technicians and managers can learn the way of the machine. They will be the employees who know how to build inherently safer plants and who, when things start to go wrong, can feel the distant shock waves building across a far-flung system. They will regard signs of trouble not as an annoyance but as rays of light, reaching into the machine's darkest corners and revealing its secrets at last. Most important, they will act before an otherwise routine day rises to disaster.

DISASTERS, CALAMITIES, AND NEAR MISSES CITED IN THE BOOK

St. Chad Cathedral: collapse
Shrewsbury, Shropshire, United Kingdom
July 9, 1788

Probably because of rotting timbers and weakened masonry in 400-year-old church. Engineer Thomas Telford had warned church elders a few weeks before that structure was unsafe.

Sultana (American steamboat): boiler explosion
Mississippi River, near Memphis, Tennessee, United States
April 27, 1865

Three of four boilers exploded while boat was transporting more than 2,000 Union ex-prisoners north. One of the boilers had been temporarily patched on trip to cover a crack. This patch was thinner than original boiler walls. Deaths: up to 1,800 (ex-prisoners, crew, and commercial passengers).

Tay Bridge (railroad bridge): collapse of mid-span during high winds
Tay River Estuary, Scotland, United Kingdom
December 28, 1879

Center of two-mile bridge had 13 high spans 88 feet over water. Bridge suffered chiefly from failure to allow for effect of strong side winds on center spans. Wind gusted to 70 mph as train crossed, causing shift of train's weight off centerline of bridge. Support columns buckled under off-center load. Chief lesson was to build structures for highest likely winds while carrying full load. Related to "P-delta" problem in steel-framed skyscrapers. Deaths: 75.

DuPont dynamite factory: explosion
Near Gibbstown, New Jersey, United States
March 18, 1884

Runaway reaction of one ton of nitroglycerine in batch nitrator. Exploded despite being dumped in "drowning tank." Blast and debris killed du Pont family member. Deaths: 5 (men on scene).

Maine (American battleship): explosion
Moored in harbor at Havana, Cuba
February 15, 1898

At time, public accepted sabotage by underwater mine as cause, but investigation in 1970s organized by Admiral Hyman Rickover concluded that triggering event was spontaneous combustion of bituminous coal in bunker (a known problem among coal-fired ships), which ignited gunpowder in magazine across bulkhead from coal bunker. Deaths: 260 (crew).

Illinois Central Railroad: train wreck
Near Vaughn, Mississippi, United States
April 30, 1900

Origin of "Casey Jones" ballad. Passenger train engineer John Luther Jones was on cannonball run, hitting speeds up to 100 mph to make up lost time. Arrived near scene of stalled train at 50–75 mph and could not stop in time. Fireman jumped clear but Jones stayed to brake train and was killed by wood fragments after locomotive rammed rear of train protruding from siding. Deaths: 1.

Quebec Bridge: collapse during construction
St. Lawrence River, Canada
August 29, 1907

Long delay, followed by bond issuance, forced completion of bridge on rapid schedule. Bridge was lengthened but design did not adjust for stresses of additional weight from change. Chief engineer did not inspect site in critical last stages. Alarmed by reports of components bending during construction, engineer sent telegram to halt work but south arm collapsed before message arrived. Deaths: 75 (workers).

Liberté (French battleship): fire and explosion in ammunition magazine
Harbor at Toulon, France
September 25, 1911

Poudre B smokeless powder for naval guns, stored in forward magazine, had been chemically degrading. Spontaneously ignited while ship was at anchor. Crew attempted to flood compartment with seawater but could not reach flooding valves due to fires and explosions. Deaths: 285 (crew and rescuers).

Titanic (British passenger liner): sunk after starboard side grazed
    iceberg
Atlantic Ocean
April 15, 1912

Sideswipe with iceberg opened holes along riveted plate seams; poor quality of rivet metal (i.e., high proportion of slag) may have aggravated damage. Total opening in hull was minimal (12 square feet) but allowed flooding of first five compartments. This weighted bow enough to bring top of sixth watertight compartment below water level, producing a chain reaction further aft with other watertight compartments. Deaths: 1,513 (crew and passengers).

Mont Blanc (French freighter): explosion
Harbor at Halifax, Nova Scotia, Canada
December 6, 1917

Following misunderstood signals, Belgian relief ship *Imo* rammed the *Mont Blanc*, which was carrying 2,500 tons of explosives (picric acid and TNT) for war effort. Sparks from collision ignited cargo of flammable liquids and explosives; blast destroyed much of north Halifax. Deaths: 2,000 (crew and Halifax residents).

BASF fertilizer factory: explosion of storage pile
Oppau, Germany
May 21, 1921

Workers at nitrate plant were blasting material loose from stockpiles of mixed ammonium nitrate and ammonium sulfate, which were thickly crusted from weathering. Workers had used this procedure over 15,000 times previously without incident. Pile exploded, destroying factory. Deaths: 560 (workers and town residents).

R.101 (British dirigible): crashed in storm
Near Beauvais, France
October 4, 1930

Airship was known by staff at Royal Airship Works to have poor performance from excess weight and leaky gasbags. Ordered out on first trip by air minister who wanted to meet schedule requirements. Crashed at low speed into forested ridge, possibly due to downdraft in windstorm. Hydrogen and envelope burned quickly. Deaths: 48 (crew and passengers).

New London Independent School: gas explosion
New London, Texas, United States
March 18, 1937

School district was heating its large school building with untreated oil field gas, purchased at a discount. Gas leaked in basement, undetected, until ignited by electrical equipment in workshop on main floor. Deaths: 298 (students and teachers).

Thetis (British submarine): sunk during sea trials
Off Welsh coast, Irish Sea
June 1, 1939

Submarine would not submerge. While checking water ballast in torpedo room, crew accidentally flooded two forward compartments. Submarine sunk and grounded at 140 feet. Suffering from poisoned air, most of crew could not manage an escape in time. Air supply lasted half its normal period because people on board totaled twice usual number. Deaths: 99 (regular crew and extra passengers).

Schenectady (American tanker): structural failure
Kaiser Swan Island Yard, Portland, Oregon, United States
January 16, 1943

Tanker broke in half while floating at dock after sea trials. Was extreme case of cracking problem among steel hulls of American freighters and tankers built during World War II. Analysis of many failures (13 percent of 2,580 ships showed serious cracks) showed most were high-speed fractures starting at notches or holes with sharp corners, such as hatch openings. Stress was concentrating at such corners and notches.

Texas City docks: explosions of two fertilizer-carrying ships
Texas City, Texas, United States
April 16, 1947

Ship *Grandcamp* held 2,200 tons of bagged ammonium nitrate and caught fire at docks. Crew tried to contain fire by sealing holds and flooding space with steam. Chemical heated first to melting point, then to a runaway reaction producing nitrous oxide. Cargo exploded at temperature of about 850°F. Second fertilizer ship, *High Flyer*, caught fire and exploded early next morning. Deaths: 468 (workers, firefighters, and bystanders).

Comet 1 (airliner): two aircraft exploded in midair in separate incidents
Mediterranean Sea
January 10, 1954, and April 8, 1954

Aluminum skin of first jet airliner suffered fatigue cracking from pressurization-depressurization cycles, though manufacturer had initially pressure-tested fuselage severely. On the only aircraft that could be recovered for study, analysis showed fracture started at corner of navigation window. Deaths: 56 (passengers and crew).

Sargo (submarine): fire in rear compartment
Pearl Harbor, Hawaii, United States
June 15, 1960

The *Sargo* was loading liquid oxygen from a tanker truck. Malfunction in submarine's stern ignited violent oxygen-enriched fire that could be extinguished only by sinking vessel at stern. Deaths: 1.

Legion Field Stadium: discovery of structural weakness in new upper deck
Birmingham, Alabama, United States
October 21, 1960

Engineer for structural steel fabricator noticed that wind bracing for stadium's new steel upper deck seemed loose. His inquiry triggered a reexamination of deck (capacity 8,632) one week before occupancy. Calculations showed design errors that put deck, if fully loaded, in danger of collapsing onto lower concrete stands. City decided not to use deck for college football game, delaying use until thorough repairs and confirmation of safety.

Minuteman 1 ICBM: deployment without full safeguards during national emergency
Malmstrom Air Force, Montana, United States
October 1962

During Cuban Missile Crisis (October 14–28) air force and consultant crews hurried to deploy a single "flight" of up to 10 new Minuteman missiles ahead of schedule. No safety interlocks or authorization-code requirements would have prevented accidental launch, because single launch control center lacked cross-checks from other centers. According to Scott Sagan research, crew kept necessary electronics for launch close to hand, instead of in a separate guarded vault, as publicly stated.

Thresher **(American submarine): sunk**
Atlantic Ocean
April 10, 1963

During sea trials at 1,300 feet, silver-brazed joint in seawater piping probably broke and sprayed salt water onto electric controls. This caused reactor to shut down, and submarine could not return to surface before flooding sunk it. Crew attempted emergency blow, but as tests revealed later, particulate strainers in air system may have been blocked by rapid ice formation from Venturi effect during emergency blow. Deaths: 129 (crew and contractor personnel).

Baldwin Hills Dam: leak followed by partial collapse and flood
Los Angeles, California, United States
December 14, 1963

Nineteen-acre lake began draining out underneath 160-foot-high earth and concrete dam, because of earth settlement at foundation. Inspector noticed widening hole, and efforts to plug hole failed. Police evacuated area downstream before dam broke. Deaths: 5.

Canadian–United States Eastern Interconnection (CANUSE):
    blackout lasting up to 13 hours
Northeast United States
November 9, 1965

Relay's failure at Canadian power plant on Niagara River caused transmission line to stop sending power into Canada; remaining three lines overloaded, shut down, and sent 1,500 megawatts of power into United States. Automatic controls began breaking region into islands, with many utilities shutting down automatically to protect generators and transformers. Many generators took hours to resume operation. Thirty million people lost power.

Apollo 1 (American spacecraft): oxygen-enriched fire in command
    module during ground testing
Kennedy Space Center, Florida, United States
January 27, 1967

Ground tests of capsule required astronauts to be inside sealed capsule pressurized with 16.7 psi, pure-oxygen atmosphere. Capsule interior held large quantities of flammable plastics. Electrical short (possibly from scraped wires near floor) ignited fire and fumes killed astronauts before rescuers could open capsule. Deaths: 3.

Hungarian Carbonic Acid Producing Company (CO_2 plant):
    explosion of cryogenic tank
Répcelak, Hungary
January 2, 1969

Liquid CO_2 storage tank exploded, possibly due to overfilling. This destroyed another tank and their fragments punctured a third tank, which took off from its base and smashed into a laboratory, freezing five people with −108°F liquid.

Apollo 13 (American spacecraft): overheating of Oxygen Tank
    Number 2 in service module external bay
Earth-Moon trajectory, 55 hours into mission
April 13, 1970
    Liquid oxygen tank had been slightly damaged at factory, shaking internal plumbing loose. Second problem with tank was thermostat rated for less voltage than rest of electrical system. Because of earlier damage tank would not empty during systems test at Kennedy Space Center, so technicians used built-in tank heater to vaporize liquid. Inside tank, power fused thermostat shut and burned off wire insulation. When during mission astronauts tried to use heater, tank top failed. Damage emptied both oxygen tanks.

American Airlines DC-10 airliner: cargo-door blowout in flight,
    damage to controls
Windsor, Ontario, Canada
June 12, 1972
    Airplane left with rear cargo door not fully locked. Air pressure differential blew out door at 12,000 feet, causing rear cabin floor to collapse into cargo comparment; this jammed control cables to elevator, rudder, and rear engine. Pilot had trained in simulator on using engines to steer with and brought aircraft back with passengers receiving only minor injuries. Precursor was cargo-door blowout during testing of first fuselage at McDonnell-Douglas factory.

Eastern Airlines L-1011 airliner: crash into swamp at night
Everglades, Florida, United States
December 29, 1972
    While attempting to confirm whether nose gear was down and locked for landing, crew put jet into autopilot for 2,000-foot altitude hold near Miami. Autopilot's control over altitude switched off after a crew member

bumped control column, allowing plane to begin a descent. Crew did not notice problem in time, over dark terrain. Deaths: 101 (crew and passengers).

Turkish Airlines DC-10 airliner: crash into forest after cargo-door blowout and loss of controls
Ermenonville, France
March 3, 1974

Rear cargo door was not locked and blew out while fully loaded aircraft was climbing out from Paris. Sudden pressure differential caused cabin floor to collapse, jamming cables to tail surfaces and rear engine. Aircraft went into dive and crashed into forest at high speed. Investigation found that aircraft had not received cargo-door safety-upgrade work that manufacturer's paperwork said it had. Deaths: 346.

Banqiao and Shimantan Dams: chain-reaction failure of dams in typhoon
Henan Province, China
August 7, 1975

Disaster was triggered in central China by very heavy rainfall, aggravated by construction and geologic problems with dams. Upstream structures failed, sending floodwaters that lower dams could not handle. Total of 62 dams collapsed in one night. Deaths: officially estimated at 26,000, possibly much higher. Information first revealed in 1995 Human Rights Watch report.

Citicorp Center (skyscraper): discovery of structural problem
New York, New York, United States
1978

Post-construction analysis by building's engineering consultant indicated that this 914-foot-tall structure was vulnerable to falling over in case of extreme but foreseeable hurricane-force winds coming from diagonal directions. Engineer notified owner, triggering urgent reinforcement of structure before fall hurricane season.

Hartford Civic Center Coliseum: collapse of space-frame roof during snowstorm
Hartford, Connecticut, United States
January 18, 1978

During construction of arena, workers reported that metal structure supporting 300-by-360-foot roof was deflecting 50 percent more than planned and some panels would not fit as designed. Arena went into service, anyway, and roof collapsed five hours after completion of a basketball game, following a four-inch snowfall. City's consultant concluded roof was built too weak.

Three Mile Island Unit 2 (nuclear reactor): partial meltdown of reactor core
Middletown, Pennsylvania, United States
March 28, 1979

While reactor was running at 97 percent of full power, maintenance work accidentally caused valves to close off secondary coolant loop. Heat trapped in reactor coolant loop caused pilot operated relief valve to vent steam but valve did not re-close (for reasons still unknown). Stuck-open relief valve caused instrument readings on coolant level to rise, so operators cut back on emergency cooling. Escape of reactor coolant continued for 2 hours 20 minutes, melting upper core, until newly arrived supervisor detected leak.

National air defense system: false alert of Soviet attack
North America
November 9, 1979

Electronic malfunction inserted information from attack-simulation training tape onto display screens in four command centers, with no indication that massive Soviet missile attack being indicated was nonexistent. Cross-checks with radar surveillance sites showed no attack was under way, and threat assessment conference ended after six minutes.

Main Mirror, Hubble Space Telescope: spherical aberration error
Perkin-Elmer Corporation, Danbury, Connecticut, United States
1980–1982

Optical manufacturer promised to grind and polish mirror blank to near-perfection but relied solely on one instrument, called a reflective null corrector, to determine curvature of glass. Assembly process for null corrector went wrong and technicians added washers to make their adjustments work. Result of maladjusted null corrector was to shape outer rim area of mirror to wrong curve. First repair occurred in December 1993.

Alexander Keilland (floating platform to house offshore workers): capsized during storm
North Sea
March 27, 1980

Platform was built with five vertical legs, cross-braced by struts. Work at shipyard left one leg with small cracks near opening for hydrophone fitting. Fractures in steel lengthened under extended battering by waves. Leg broke away during storm, putting platform off balance and starting capsize. Deaths: 123.

National air defense system: false alert of Soviet attack
North America
June 3, 1980

Computer chip failed, causing displays to show varying number of Soviet missiles in flight. Cross-checks showed information was false. Problem repeated three days later.

Hyatt Regency Hotel: collapse of two skywalks during "tea dance"
Kansas City, Missouri, United States
July 17, 1981

Original design of skywalks for high lobby changed during construction to simplify method of suspending structures from steel rods. Result was that fourth-level walkway had to bear weight of second-level walkway suspended below. During social event, connection of steel rods to upper skywalk failed, sending crowded skywalks onto more guests below. Deaths: 114.

Ocean Ranger (offshore floating drill rig): sunk in storm
Grand Banks, Atlantic Ocean
February 15, 1982

Storm wave broke out port-light window into ballast control room of floating "semisubmersible" drill rig. Seawater shorted out electrical gear and valves began operating randomly. Rig tilted and crew tried to restore stability, not knowing that when rig was at an angle pumps would do opposite of what they expected. As tilt worsened, waves sent water into openings in upper deck. Two of three lifeboats were destroyed in evacuation. Deaths: 84 (entire crew).

Union Carbide (India) Ltd. plant: release of highly toxic chemicals into community
Bhopal, Mahdya Pradesh, India
December 3, 1984

According to Indian government, factory producing highly toxic methyl isocyanate (MIC) for herbicides was poorly maintained, particularly safety equipment. Storage tank vented tons of MIC after water from pipe-washing job accidentally got into tank, causing violent reaction that remaining safeties could not handle. Union Carbide claims worker sabotage put water into tank; third theory says workers mistakenly put water into tank to "cool it off." Deaths: officially estimated 7,000.

Galaxy Airlines Electra II airliner: crash during emergency return to airport
Reno, Nevada, United States
January 21, 1985

Because of distraction by an interruption, ground crew did not fully secure access door on wing for air-start equipment door on airliner wing. Door came loose and caused severe buffeting in flight, from disruption of airstream over wing. Not knowing cause, pilots requested return to airport, believing aircraft was in structural danger. Pilots reduced engine power and aircraft crashed when airspeed dropped too low. Deaths: 70.

British Airtours 737 airliner: aborted takeoff and fire
Manchester, United Kingdom
August 22, 1985

Combustion "can" on left-hand jet engine flew apart on takeoff roll and one fragment crashed through an access panel, allowing fuel to pour out. After aircraft stopped, fire on left side, pushed by breeze, burned through fuselage. Cabin filled with toxic smoke and fumes, making evacuation extremely difficult. Deaths: 55.

Space shuttle *Challenger* (American spacecraft): midair breakup
Near Kennedy Space Center, Florida, United States
January 28, 1986

At time of liftoff, hot gas penetrated seal on right-hand booster, near lower strut connecting booster to external tank. Small hole sealed, then reopened later. Flame caused strut to fail 73 seconds into flight. Booster hinged around remaining upper struts, crashing into external tank. Tank failure threw orbiter sideways into supersonic slipstream, breaking it apart. Any survivors in crew compartment died on impact with water. Deaths: 7.

V. I. Lenin Chernobyl Nuclear Power Station, Reactor 4: explosions
Near Pripyat, Ukraine, Russia
April 26, 1986

Crew was under orders during scheduled shutdown to measure how long residual energy could produce electricity after nuclear reaction was stopped. Crew continued, though reactor proved unstable at low power. When operators tried to use control rods, graphite tips of rods displaced water from channels and brought power level to 100 times maximum allowed, causing two explosions. Government evacuated 135,000 people from 1,000-square-mile area. Deaths: estimated 5,000–10,000 (mostly cleanup workers).

Aloha Airlines 737 airliner: blowout of upper fuselage skin
Maui, Hawaii, United States
April 28, 1988

Aluminum corrosion, adhesive weakening, and fatigue from use of 19-year-old aircraft for short-haul flights caused fracture and explosive blowout of skin at 24,000 feet altitude. Made hole 18 feet in length. Airflow sucked one flight attendant out. Pilot retained control and landed safely. Deaths: 1.

Pacific Engineering & Production Company of Nevada: fire and explosion of 4,500 tons of ammonium perchlorate
Henderson, Nevada, United States
May 4, 1988

Factory near Reno made oxidizer chemical for solid-rocket motors and had unusually large stocks on hand in 1988 because of suspension of space shuttle flights after the *Challenger* disaster. Approximately 4,500 tons of ammonium perchlorate stocks burned, then exploded. Fire department blamed welding torch for trigger; company blamed leaking gas pipeline. Deaths: 2 (managers on-site).

Piper Alpha (offshore drilling rig): explosion and fire
North Sea
July 6, 1988

Offshore platform processed large volumes of natural gas from other rigs via pipes. Daytime crew was repairing gas-condensate pump but didn't finish. Repairmen relayed verbal message about pump work but next shift's workers turned pump on anyway. Temporary seal on pump failed and fire broke open main "risers" from other rigs, causing much larger fire that trapped many crewmen in crew quarters, high over ocean with no escape route. Deaths: 167 (crew and rescue workers).

British Midlands 737 airliner: crash
Near Kegworth, Leicestershire, United Kingdom
January 8, 1989

Flight crew concluded from noise and vibration, and from use of throttles, that right-hand engine on twin-engined airliner was about to fail, so crew shut down that engine and relied on left-hand engine. In fact, left-hand engine had been failing due to ingestion of broken fan blade and lost all power short of airport. Crew did not have time to restart right-hand engine and crashed during approach. Deaths: 47.

Exxon Valdez (American oil tanker): grounded
Bligh Reef, Prince William Sound, Alaska, United States
March 24, 1989

Outbound tanker left marked channel in attempt to avoid pack ice, with captain leaving bridge before difficult maneuver approached. Tanker failed to make turn in time and grounded on Bligh Reef. Ship released 11 million gallons of crude oil. Ecological damage was aggravated by slow spill response time.

British Airways BAC 1–11 airliner: ejection of pilot through windscreen
Near Didcot, Oxfordshire, England
June 10, 1990

Supervisor replaced windscreen on nose of airliner during overnight maintenance shift, but 84 of 90 bolts he installed were too narrow, leaving only six holding well. Windscreen blew out at 17,000 feet and escaping cabin air pulled captain out through hole, leaving him pinned back against window frame as flight attendants held onto his feet. Captain survived 18-minute flight back to airport in this position.

Lauda Air 767 airliner: loss of control and crash
Departing Bangkok, Thailand
May 26, 1991

Electrical and mechanical problems caused deployment of thrust reverser on left-hand engine, during climb-out. Computerized safety interlock did not stow reverser as intended. Reverser disrupted lift on left side and threw aircraft into immediate left diving turn. Dive speed reached .99 Mach, possibly passing speed of sound, before structural midair breakup. Deaths: 223.

Royal Majesty (cruise ship): grounded on shoal
East of Nantucket Island, Massachusetts, United States
June 10, 1995

Ship grounded after failure in ship's integrated navigation system, unnoticed by bridge crew. Global Positioning System antenna cables had come loose, and navigation had shifted to a dead-reckoning mode that only maintained last heading. Also, depth alarm had been wrongly adjusted to warn crew only when zero depth of water remained under keel.

ValuJet DC-9 airliner: crash following fire in cargo compartment
Everglades, Florida, United States
May 1, 1996

Maintenance contractor accumulated dozens of expired chemical oxygen canisters following work on customer's jets. Mechanics taped canisters, and contractor arranged for shipment to customer's headquarters as freight on another airliner. NTSB reports concluded that canisters triggered early in flight and heat of chemical reaction caused packing materials to catch fire. Tires also being shipped as cargo caught fire. Oxygen-enriched fire burned through compartment liner and cut off flight crew's engine controls while aircraft was attempting to return to Miami. Deaths: 110 (crew and passengers).

AeroPeru 757 airliner: crash into ocean following loss of control
Pacific Ocean, off Peru
October 2, 1996

Maintenance crew washed plane and left pitot tubes taped over. Because airspeed and altitude indicators needed air from tubes to function, erratic readings confused crew during nighttime flight. Aircaft crashed into ocean after 30 minutes. Deaths: 68 (crew and passengers).

Instituto Galeazzi hyperbaric therapy chamber: fire
Milan, Italy
October 31, 1997

Fire in pressurized therapy chamber killed patients and nurse. Patients were breathing pure oxygen through masks. Warnings went out afterward that patients must not only be warned about bringing in any sources of ignition but need to be searched as well. Deaths: 11.

Four Times Square (skyscraper): collapse of scaffold and hoist mast
on construction site
New York, New York, United States
July 21, 1998

Scaffold tower buckled at twenty-first floor, dropping upper 300 feet of mast onto streets and nearby building roofs. Fifty-foot section of mast crashed through roof of residence hotel. Deaths: 1.

Minneapolis Police van: crash into crowd during holiday parade
Minneapolis, Minnesota, United States
December 4, 1998

Federal investigation concluded that while moving a police van at a holiday parade, police officer must have pressed gas pedal instead of brake when shifting out of park. Van took off at full throttle and crashed into a crowd, stopping only after it hit a building column. A common police department modification to van (intended to make taillights flash) had unintentionally deactivated electric "shift lock" device installed by manufacturer to prevent such sudden-acceleration accidents. Deaths: 2.

Mars Climate Orbiter **(American space probe): loss of spacecraft**
Initial flyby around Mars
September 23, 1999

Craft came too close on initial aerobraking approach to Mars (approximately 35 miles instead of 140 miles) because of small errors accumulating over months, due to programming error in software that had been running on Earth-based computer. Craft either broke up or went into solar orbit. Computer, which processed information that orbiter transmitted about its correctional thruster burns, should have reported data in metric units but used English units instead.

JCO Tokai Works Conversion Test Facility uranium processing plant:
    accidental critical mass
Tokaimura, Japan
September 30, 1999

Workers had job of chemically preparing uranium fuel for later use in reactors. Decided to shorten process by transferring dissolved uranium via buckets directly to precipitator, bypassing intermediate steps. Sufficient uranium solution gathered in precipitator to cause a critical mass, a low-grade nuclear reactor, that irradiated three workers. Deaths: 1.

Sunjet Aviation Learjet: crashed while carrying golfer Payne Stewart
    and party
Near Aberdeen, South Dakota, United States

October 25, 1999

Cabin pressurization system failed above 30,000 feet over northern Florida for reasons unknown, causing rapid loss of consciousness among pilots and passengers. Airplane continued on autopilot until ran out of fuel over South Dakota and crashed. Airplane had lost partial cabin pressure before. Deaths: 6.

Mars Polar Lander (American space probe): loss of spacecraft
Mars surface
December 3, 1999

Programming mistake fooled lander into cutting off its braking thruster too soon, when it was still 130 feet above surface. Reason was jolt from deployment of landing legs, which sensors misinterpreted as touchdown on surface. Problem was not noticed on Earth during checkout tests because of a wiring error when tests were conducted; wiring was fixed later but relevant tests were not redone.

Air France Concorde supersonic transport: crash
Departing Charles de Gaulle Airport, Paris, France
July 25, 2000

Titanium debris on runway caused tire to burst on takeoff. Ten-pound fragment of tire hit left wing tank and impact caused shock wave in fuel that broke out section of tank wall from inside. Rapid fuel leak (26 gallons per second) opened forward of engine intakes. Fuel vapor and flame into left engines caused loss of power. Deaths: 113 (passengers, crew, and victims on ground).

LIST OF KEY SOURCES*

INTRODUCTION

Ansoff, H. Igor. *Strategic Management*. New York: John Wiley & Sons, 1979.

Bignell, Victor. *Understanding System Failures*. Manchester, England: Manchester University Press, 1984.

Came, Barry. "Takeoff to Disaster." *Maclean's*, August 7, 2000, 20.

Canning, John, ed. *Great Disasters: Catastrophes of the Twentieth Century*. Norwalk, Conn.: Longmeadow Press, 1976.

Churchill, Winston. *The World Crisis*. New York: Charles Scribner's Sons, 1923.

Court of Inquiry. *Civil Aircraft Accident: Comet G-ALYP on 10 January 1954 and Comet G-ALYY on 8 April 1954. London: Ministry of Transport and Civil Aviation*. London: Stationery Office Publications Centre, 1955.

Davis, Lee. *Man-Made Catastrophes*. New York: Facts on File, 1993.

"Explosions—At Old and New Plants—Concern the Industry." *Power*, September–October 1999, 13.

Ford, Daniel F. *The Button*. New York: Simon & Schuster, 1985.

Fulghum, David A. "USAF Maps Solution to Rapidly Aging Parts." *Aviation Week & Space Technology*, June 12, 2000, 51.

---

*Note: A complete bibliographic list is posted at www.invitingdisaster.com.

Gold, Michael. "Who Pulled the Plug on Lake Peigneur?" *Science 81*, November 1981, 56.

Learmount, David. "Fresh Crash Data Offer Hope for Concorde Recertification." *Flight International*, January 16, 2001, 11.

Marino, Sal. "Murphy's Laws Are Poor Excuses for Mismanagement." *Industry Week*, October 4, 1999, 20.

"Overreliance on Automation Led to Cruise Ship Grounding." *The Mobility Forum*, May–June 1999, 26.

Perreault, William D., and Anthony Vandyk. "Did the Jet Age Come Too Soon?" *Life*, January 25, 1954, 54.

Post, Nadine. "Causes Probed as Crippled Hoist Comes Slowly Down." *Engineering News-Record*, August 10, 1998, 10.

Qing, Dai. *The River Dragon Has Come!* Armonk, N.Y.: M. E. Sharpe, 1998.

Rasmussen, Jens, Keith Duncan, and Jacques LePlat. *New Technology and Human Error.* New York: John Wiley & Sons, 1987.

Roush, Wade. "Learning from Technological Disasters." *Technology Review*, September 1993, 51.

Schlager, Neil. *When Technology Fails.* Detroit: Gale Research, 1993.

Schwebel, Jim. 2000. Interview by author. 16 June.

Singleton, W. T. *Man-Machine Systems.* Harmondsworth, England: Penguin Education, 1974.

Sparaco, Pierre. "Europe Embarks on $11-Billion A380 Gamble," *Aviation Week & Space Technology*, January 1, 2001, 22.

Turner, Barry A. *Man-Made Disasters.* London: Wykeham Publications, 1997.

Unsworth, Edwin. "Prospect of Huge Ships Concerns Insurers." *Business Insurance*, September 27, 1999, 16.

U.S. National Transportation Safety Board. *Galaxy Airlines Inc., Lockheed Electra-L-188C, N5532, Reno, Nevada, January 21, 1985.* Washington, D.C.: U.S. General Printing Office, 1986.

Voros, M., and Gy Honti. "Explosion of a Liquid CO_2 Storage Vessel in a Carbon Dioxide Plant," in Buschmann, C. H., ed. *Loss Prevention and Safety Promotion in the Process Industries.* Amsterdam: Elsevier Scientific Publishing, 1974.

Whyte, R. R., ed. *Engineering Progress Through Trouble.* London: Institution of Mechanical Engineers, 1975.

CHAPTER 1

Cox, R. F., and M. H. Walter. *Offshore Safety and Reliability*. London: Elsevier Applied Science, 1991.

Gerber, Marius. 2000. Personal communication with author. 13 October.

Gibb, Bob. "*Odyssey*—A Fatal Error of Judgment." *Lloyd's List*, November 9, 1991, 3.

House, J. D. *But Who Cares Now?* St. John's, Newfoundland: Breakwater Books, 1987.

Hylund, Richard. 2000. Interview by author. 24 August.

Joyce, Randolph. "The Cruel Sea." *Maclean's*, March 1, 1982, 26.

LeBlanc, Leonard. "Tracing the Causes of Rig Mishaps." *Offshore*, March 1981, 51.

Milgram, Jerry. 2000. Telephone interviews by author. November.

Muggeridge, Derek. 2000. Telephone interview by author. 3 October.

O'Flaherty, Patrick. "Salvaging the Truth: An Investigation into the Sinking of the *Ocean Ranger*." *Canadian Forum*, January 1987, 16.

Rathbun, Robert. 2000. Interview by author. 20 May.

Royal Commission on the *Ocean Ranger* Marine Disaster. *Report One: The Loss of the Semisubmersible Drill Rig Ocean Ranger and Its Crew*. St. John's, Newfoundland: Royal Commission on the *Ocean Ranger* Marine Disaster, 1984.

Savage, Diane. 2000. Interview by author. 20 May.

Turner, Lindsay. 2000. Interview by author. 24 August.

U.S. Coast Guard Marine Board of Investigation. *Report Concerning the MODU Ocean Ranger, O.N. 615641*. Washington, D.C.: U.S. General Printing Office, 1982.

U.S. National Transportation Safety Board. *Capsizing and Sinking of the U.S. Mobile Offshore Drilling Unit Ocean Ranger, February 15, 1982*. Washington, D.C.: U.S. General Printing Office, 1983.

Winsor, Gordon. 2001. Telephone interview by author. 31 January.

Wright, Colin. "*Ocean Odyssey* Gas Explosion Report Critical of Operator." *The Oilman*, December 1991, 63.

Zoellner, Tom. "Oil and Water." *American Heritage of Invention & Technology*, (Fall 2000): 44.

CHAPTER 2

"Atomic Miracles Are Nearer Now." *Popular Science*, June 1954, 151.

Boyle, R. H. "Nukes Are in Hot Water." *Sports Illustrated*, January 20, 1969, 24.

Clarfield, Gerard H., and William M. Wiecek. *Nuclear America: Military and Civilian Nuclear Power in the United States 1940–1980*. New York: Harper & Row, 1984.

Corddry, Charles. "What Is Happening to Our Aircraft Nuclear Propulsion Program?" *Flying*, April 1959, 32.

Dietz, David. *Atomic Energy in the Coming Era*. New York: Dodd, Mead, 1945.

Duffy, L. P., E. E. Kintner, R. H. Fillnow, and J. W. Fisch. "The Three Mile Island Accident and Recovery." *Nuclear Energy*, August 1986, 199.

"Fiery Setting for A-Power: A New Nuclear Era Shapes Up Underground at Shippingport." *Life*, October 1, 1956, 130.

Ford, Daniel F. *Three Mile Island: Thirty Minutes to Meltdown*. New York: Viking, 1982.

Gray, Mike. *The Warning*. New York: W. W. Norton, 1982.

Greenlee, Jason M., David Thomas, and Paul Gleason. "From Cosmos to Chaos: Déjà Vu or Vu Jade?" *Wildfire*, March 1996.

Hagen, E. W. "Common Mode Failure." *Nuclear Safety*, March–April 1980, 184.

Labich, Kenneth. "Elite Teams Get the Job Done." *Fortune*, February 19, 1996, 90.

Lanir, Zvi. *Fundamental Surprise: The National Intelligence Crisis*. Tel Aviv, Israel: Center for Strategic Studies, 1983.

Lanouette, William. "Atomic Energy, 1945–1985." *Wilson Quarterly* (Winter 1985): 90.

LaPorte, Todd R., and Paula M. Consolini. "Working in Practice But Not in Theory: Theoretical Challenges of High Reliability Organizations." *Journal of Public Administration Research and Theory* (January 1991): 19.

Laurence, William L. "Peaceful Uses for the H-Bomb." *Science Digest*, October 1958, 18.

MacPherson, Malcolm. *The Black Box: All New Cockpit Voice Recorder Accounts of In-Flight Accidents.* New York: William Morrow, 1998.

Mehler, Brian. 2000. Telephone interview by author. 27 August.

"Nuclear Ditch-Digging." *Business Week*, December 21, 1963, 84.

"The Peaceful Atom: Friend or Foe?" *Time*, January 19, 1970, 42.

Perrow, Charles. *Normal Accidents: Living with High-Risk Technologies.* Princeton, N.J.: Princeton University Press, 1999.

President's Commission on the Accident at Three Mile Island. *The Report of the President's Commission on the Accident at Three Mile Island. The Need for Change: The Legacy of TMI.* Washington, D.C.: U.S. General Printing Office, 1979.

Rochlin, Gene I., Todd R. LaPorte, and K. H. Roberts. "The Self-Designing High-Reliability Organization: Aircraft Carrier Flight Operations at Sea." *Naval War College Review* (Autumn 1987): 76.

Rubin, A. M., and E. Beckjord. "Three Mile Island—New Findings 15 Years after the Accident." *Nuclear Safety* 35, no. 2 (December 1994).

Sills, David L. H., C. P. Wolf, and Vivien B. Shelanski. *Accident at Three Mile Island.* Boulder, Colo.: Westview Press, 1982.

Stephens, Mark. *Three Mile Island.* New York: Random House, 1980.

Stokley, James. *The New World of the Atom.* New York: Ives Washburn, 1970.

"Three Mile Island Crew Struggled with the Impossible." *The Washington Star*, June 10, 1979, 1.

U.S. Congress, Subcommittee on Nuclear Regulation, Committee on Environment and Public Works, U.S. Senate. *Nuclear Accident and Recovery at Three Mile Island.* Washington, D.C.: U.S. General Printing Office, 1980.

U.S. Nuclear Regulatory Commission Special Inquiry Group. *Three Mile Island: A Report to the Commissioners and the Public.* Washington, D.C.: U.S. General Printing Office, 1980.

Webb, Richard E. *The Accident Hazards of Nuclear Power Plants.* Amherst, Mass.: University of Massachusetts Press, 1976.

Weick, Karl E. "The Collapse of Sensemaking in Organizations: The Mann Gulch Disaster." *Administrative Science Quarterly* (December 1993): 628.

CHAPTER 3

Bell, Trudy E. "The Space Shuttle: A Case of Subjective Engineering." *IEEE Spectrum* (June 1989): 42.

Biddle, Wayne. "Two Faces of Catastrophe." *Air & Space*, September 1990, 46.

Boisjoly, Roger M. "The *Challenger* Disaster: Moral Responsibility and the Working Engineer," in *Ethical Issues in Engineering*, edited by Deborah G. Johnson. Englewood Cliffs, N.J.: Prentice Hall, 1991.

Casamayou, Maureen Hogan. *Bureaucracy in Crisis: Three Mile Island, the Shuttle Challenger, and Risk Assessment*. Boulder, Colo.: Westview Press, 1993.

Duncan, Francis. *Rickover and the Nuclear Navy: The Discipline of Technology*. Annapolis: Naval Institute Press, 1990.

Foecke, Timothy. 2001. Telephone interview by author. 15 January.

Higham, Robin. *The British Rigid Airship, 1908–1931: A Study in Weapons Policy*. London: G. T. Foulis, 1961.

Hoare, Sir Samuel. *Empire of the Air*. London: Collins, 1957.

Jenkins, Dennis R. *Space Shuttle: The History of Developing the National Space Transportation System*. Marceline, Mo.: Walsworth Publishing, 1993.

Leasor, James. *The Millionth Chance: The Story of the* R.101. London: Hamish Hamilton, 1957.

Lewis, Richard S. *Challenger: The Final Voyage*. New York: Columbia University Press, 1988.

Mark, Robert. *Experiments in Gothic Structure*. Cambridge, Mass.: MIT Press, 1987.

Masefield, Sir Peter G. *To Ride the Storm: The Story of the Airship* R.101. London: William Kimber, 1982.

President's Commission on the Space Shuttle *Challenger* Accident. *Report of the President's Commission on the Space Shuttle* Challenger *Accident*. Washington, D.C.: U.S. General Printing Office: 1986.

Ritchey, Harold. 1995. Interview by author. 4 November.

Rockwell, Theodore. *The Rickover Effect: How One Man Made a Difference*. Annapolis: Naval Institute Press, 1992.

Starbuck, William H. "Challenger: Fine-Tuning the Odds Until Something Breaks." *Journal of Management Studies* (July 4, 1988): 319.

Trento, Joseph J. *Prescription for Disaster.* New York: Crown, 1987.

Vaughan, Diane. *The* Challenger *Launch Decision: Risky Technology, Culture, and Deviance at NASA.* Chicago: University of Chicago Press, 1996.

CHAPTER 4

Aircraft Accident Investigation Committee, Ministry of Transport and Communications, Thailand. "Lauda Air Luftfahrt Aktiengesellschaft Boeing 767-300ER, Registration OE-LAV, Dan Chang District, Suphan Buri Province, Thailand, 26 May B.E. 2534." http://www.rvs.unibielefeld.de/publications/incidents/DOCS/ComAndRep/LaudaAir/LaudaRPT.html (Accessed January 10, 2001).

Blair, Clay. *Silent Victory: The U.S. Submarine War Against Japan.* (Vol. 2.) Philadelphia: Lippincott, 1975.

Capers, Robert S., and Eric Lipton. "Hubble Error." *Academy of Management Executives* (November 1993): 41.

Chaisson, Eric J. *The Hubble Wars: Astrophysics Meets Astropolitics in the Two-Billion-Dollar Struggle over the Hubble Space Telescope.* New York: HarperPerennial, 1995.

Dornheim, Michael A. "Aerospace Corp. Study Shows Limits of Faster-Better-Cheaper." *Aviation Week & Space Technology,* June 12, 2000, 47.

Gannon, Robert. *Hellions of the Deep.* University Park: Pennsylvania State University Press, 1996.

Jet Propulsion Laboratory Special Review Board. *Report of the Loss of the Mars Polar Lander and Deep Space 2 Missions.* Washington, D.C.: U.S. General Printing Office, 2000.

Mares, George Carl. *The History of the Typewriter.* London: Guilbert Pitman, 1909.

"NASA's Not Shining Moments." *Scientific American,* February 2000, 13.

Norris, Guy, and Mark Wagner. *Douglas Jetliners.* Osceola, Wis.: MBI Publishing, 1999.

Parsons, Paul. "NASA Scrambling to Surmount Orbiter Loss." *Astronomy,* January 2000, 32.

Perin, Constance. "Operating as Experimenting: Synthesizing Engineering and Scientific Values in Nuclear Power Production." *Science, Technology & Human Values* (Winter 1998): 98.

Scott, R. L. "Fuel-Melting Incident at the Fermi Reactor on October 5, 1966." *Nuclear Safety*, March–April 1971, 123.

Travis, John. "Mars Observer's Costly Solitude," *Science*, September 3, 1993, 1264.

U.S. National Aeronautics and Space Administration. *The Hubble Space Telescope Optical Systems Failure Report*. Washington, D.C.: U.S. General Printing Office: 1990.

CHAPTER 5

Barlay, Stephen. *Aircrash Detective—The Quest for Aviation Safety: An International Report*. London: Hamish Hamilton, 1969.

Bates, Warren. "Shattered Windows, Lives." *Las Vegas Review-Journal*, May 3, 1998.

Burcham, Frank. 2000. Personal communication with author. 16 August.

Campion-Smith, Bruce. *Boeing 777*. Shepperton, England: Ian Allan, 1997.

Davidson, Art. *In the Wake of the* Exxon Valdez: *The Devastating Impact of the Alaska Oil Spill*. San Francisco: Sierra Club Books, 1990.

Eddy, Paul, Elaine Potter, and Bruce Page. *Destination Disaster: From the Tri-Motor to the DC-10: The Risk of Flying*. New York: Quadrangle Press, 1976.

Faith, Nicholas. *Black Box*. London: Macmillian, 1997.

Gibson, Jeff. 2001. Personal communication with author. 15 January.

Godson, John. *The Rise and Fall of the DC-10*. New York: David McKay, 1975.

Heppenheimer, T. A. *Turbulent Skies: The History of Commercial Aviation*. New York: John Wiley & Sons, 1995.

Irving, Clive. *Wide-body: The Triumph of the 747*. New York: William Morrow, 1993.

Kahneman, Daniel, Paul Slovic, and Amos Tversky. *Judgment under Uncertainty: Heuristics and Biases*. Cambridge: Cambridge University Press, 1982.

Kruger, Justin, and David Dunning. "Unskilled and Unaware of It: How Difficulties in Recognizing One's Own Incompetence Lead to Inflated Self-Assessments." *Journal of Personality and Social Pyschology*, (December 1999): 1121.

Kruger, Justin. 2000. Interview by author. 10 August.

McCormick, Bonnie. 2000. Telephone interview by author. 24 August.

Smith, Cydya. 2000. Interview by author. 25 May.

Stafford, Edward. *The Far and the Deep*. New York: Putnam, 1967.

Stewart, Stanley. *Emergency! Crisis in the Cockpit*. Blue Ridge, Penn.: Tab Books, 1991.

Strickland, Butch. 1997. Interviews by author. January.

Stromquist, Walter. 2000. Personal communication with author. 14 January.

U.S. National Transportation Safety Board. *Abstract of Final Report, Aviation Accident near Aberdeen, South Dakota, October 25, 1999*. Washington, D.C.: U.S. General Printing Office, 2000.

Warren, C. E. T. and James Benson. *The Admiralty Regrets . . . The Story of His Majesty's Submarine "Thetis" and "Thunderbolt."* London: George G. Harrap, 1958.

Woodward, David. *Sunk! How the Great Battleships Were Lost*. London: George Allen & Unwin, 1982.

CHAPTER 6

Archer, Mike. 2000. Personal communication with author. 5 October.

Bowen, John E. "Unique Hazard Posed by Oxygen-Enriched Atmosphere." *Fire Engineering*, March 1982, 38.

Bryan, Coleman. 2000. Telephone interview by author. 21 August.

Colley, David P. "The Lessons Learned from SSN 593." *Mechanical Engineering*, February 1987, 54.

Collins, Michael. *Carrying the Fire: An Astronaut's Journey*. New York: Farrar Straus and Giroux, 1974.

Failes, Geoff. "BHP on Brink of Disaster." *Illawarra (Australia) Mercury*, November 25, 1997, 1.

Feith, Greg. 2000. Telephone interview by author. 17 October.

Gayle, John B. "Explosions Involving Liquid Oxygen and Asphalt." *Fire Journal* (May 1973): 12.

Greissel, Mike. "The Power of Oxygen." *New Steel*, April 2000, 24.

Heppenheimer, T. A. *Countdown: A History of Space Flight*. New York: John Wiley & Sons, 1997.

Jones, Jon. 2000. Telephone interview by author. 26 September.

Jones, Jon C., and Ronald W. Mike. "Handling Liquid Oxygen Emergencies in the Home." *Fire Command*, July 1984, 16.

Jones, Richard. 2000. Interview by author. 24 August.

Kleinberg, Eliot. "Oxygen Fuels Inferno in ValuJet Test." *Palm Beach Post*, November 20, 1996: 1A.

Mondale, Walter. 2000. Interview by author. 15 February.

Moore, John. *The Wrong Stuff: Flying on the Edge of Disaster*. Ocean, N.J.: Specialty Press, 1997.

Oberg, James. *Uncovering Soviet Disasters*. New York: Random House, 1988.

Polmar, Norman. *Death of the Thresher*. New York: Chilton Books, 1964.

Sagan, Scott. 2000. Telephone interview by author. 26 September.

Sagan, Scott. *The Limits of Safety: Organizations, Accidents, and Nuclear Weapons*. Princeton, N.J.: Princeton University Press, 1995.

Sansone, Robert. 2000. Interview by author. 24 August.

Simini, Bruno. "Fire Fuels Concerns Over Hyperbaric Oxygen Facilities." *The Lancet* (November 8, 1997): 1375.

Slayton, Donald K. *Deke! U.S. Manned Space: From Mercury to the Shuttle*. New York: Tom Doherty Associates, 1994.

Stein, Peter, and Peter Feaver. *Assuring Control of Nuclear Weapons*. Cambridge, Mass.: Center for Science and International Affairs, Harvard University, 1987.

Technical Committee on Oxygen-Enriched Atmospheres. *NFPA 53: Recommended Practice on Materials, Equipment and Systems Used in Oxygen-Enriched Atmospheres*. Quincy, Mass.: National Fire Protection Association, 1999.

U.S. Congress. Joint Committee on Atomic Energy. *The Loss of the U.S.S. Thresher*. Washington, D.C.: U.S. General Printing Office, 1965.

U.S. National Transportation Safety Board. *In-flight Fire and Impact with Terrain ValuJet Airlines Flight 592 DC-9-32, N904VJ Everglades Near*

Miami, Florida May 11, 1996. Washington, D.C.: U.S. Government Printing Office, 1997.

CHAPTER 7

Atkinson, William. "Wake Up! Fighting Fatigue in the Workplace." *Risk Management* (November 1999): 11.

Beaty, David. *Naked Pilot.* Shrewsbury, England: Airlife Publishing, 1995.

Berkun, M. M. "Performance Decrement Under Psychological Stress," *Human Factors* (February 1964): 21.

Chernousenko, V. M. *Chernobyl: Insight from the Inside.* Berlin: Springer-Verlag, 1991.

"The Explosions That Shook the World," *Science* (April 19, 1996): 352.

Franklin, Ned. "The Accident at Chernobyl." *The Chemical Engineer* (November 1996): 17.

Lovell, James. *Apollo 13.* Boston: Hougton Mifflin, 2000.

Maclean, Norman. *Young Men and Fire.* Chicago: University of Chicago Press, 1992.

McConnell, Malcolm. "Rigged for Danger." *Reader's Digest,* September 2000, 128.

Medvedev, Grigori. *The Truth About Chernobyl.* New York: Basic Books, 1991.

Selye, Hans. *Selye's Guide to Stress Research.* New York: Van Nostrand Reinhold, 1983.

Sweet, William. "Chernobyl: What Really Happened." *Technology Review* (July 1989): 43.

U.S. Nuclear Regulatory Commission. *Report on the Accident at the Chernobyl Nuclear Power Station.* Washington, D.C.: U.S. General Printing Office, 1987.

CHAPTER 8

Apollo 13 Review Board. *Report of* Apollo 13 *Review Board.* Washington, D.C.: U.S. General Printing Office, 1970.

Bower, Bruce. "Seeing Through Expert Eyes." *Science News*, July 18, 1998, 44.

Bradford, Michael. "Near-Miss Analysis Directs Safety Plan." *Business Insurance*, September 27, 1999, 3.

Cook, George. "Engineers Assume All Responsibility." *Birmingham Post*, October 24, 1961, B-1.

Feld, Jacob, and Kenneth L. Carper. *Construction Failure*. New York: John Wiley & Sons, 1997.

Kaminetzky, Dov. *Design and Construction Failures: Lessons from Forensic Investigations*. New York: McGraw-Hill, 1991.

Klein, Gary. *Sources of Power*. Cambridge: Massachusetts Institute of Technology, 1998.

Kranz, Gene. *Failure Is Not an Option*. New York: Simon & Schuster, 2000.

LeMessurier, William. 2000. Telephone interview by author. 22 May.

"Man of the Year: Leslie E. Robertson." *Engineering News-Record*, February 23, 1989, 39.

McKaig, Thomas H. *Building Failures: Case Studies in Construction and Design*. New York: McGraw-Hill, 1962.

McRoy, Wallace. 2000. Telephone interview by author. 14 August.

Morgenstern, Joe. "The Fifty-nine-Story Crisis." *New Yorker*, May 29, 1995, 45.

Novomesky, John. 2000. Interview by author. 26 August.

Petroski, Henry. *To Engineer Is Human: The Role of Failure in Successful Design*. New York: St. Martin's Press, 1985.

Robertson, Leslie. 2000. Interviews by author. May.

Ross, Steven S., ed. *Construction Failures*. New York: McGraw-Hill, 1984.

Sasser, Myron. 2000. Telephone interview by author. 1 August.

Smith, Art. 2000. Telephone interview by author. 25 September.

Swift, Ivan, and Bud Gordon. "Designer Says Computation off on Legion Field Truss." *Birmingham News*, October 23, 1961, 1.

Tipper, C. F. *The Brittle Fracture Story*. Cambridge: Cambridge University Press, 1962.

"What Really Happened to *Apollo 13*," *Space World*, October 1970, 20.

CHAPTER 9

Biasutti, G. S. *History of Accidents in the Explosive Industry.* Vevey, Switzerland: G. S. Biasutti, 1985.

Botsford, Harry. "The Most Dangerous Job in the World." *Saturday Evening Post,* January 27, 1951, 26.

Campolong, Mark. 2000. Interview by author. 28 August.

Course, A. G. *The Deep Sea Tramp.* Barre, Mass.: Barre Publishing, 1963.

Dutton, William S. *One Thousand Years of Explosives.* Philadelphia: John C. Winston, 1960.

Ed, Darryl. 2000. Interview by author. 19 May.

Feerst, Robert. 2000. Interview by author. 31 August.

Fethers, Rick. 2000. Interviews by author. July.

Halasz, Nicholas. *Nobel—A Biography of Alfred Nobel.* New York: Orion Press, 1959.

Harshaw, Jack. "Calm Cleanup in Progress Today in Aftermath of Severe Blasts." *Carthage (Missouri) News,* July 15, 1966, 1.

Johnson, Robert Erwin. *Guardians of the Sea: History of the United Coast Guard, 1915 to the Present.* Annapolis: Naval Institute Press, 1987.

Kurtgis, Michael. 2000. Interviews by author. 23 August and 1 October.

Lane, Doug. 2000. Interview by author. 23 August.

Meidl, James H. *Explosive and Toxic Hazardous Materials.* Beverly Hills: Glencoe, 1969.

Parker, Brian. 2000. Interview by author. 19 May.

Stephens, Hugh. *The Texas City Disaster.* Austin: University of Texas Press, 1997.

Trout, Tom. 2000. Interview by author. 6 July.

Van Gelder, Arthur Pine, and Hugo Schlatter. *History of the Explosives Industry in America.* New York: Columbia University Press, 1927.

Whitesell, Jesse. 2000. Interview by author. 6 July.

Wilson, Elizabeth. "Processing Errors Led to Nuclear Accident." *Chemical & Engineering News,* October 11, 1999, 19.

CHAPTER 10

Air Accidents Investigation Branch. *Report on the Accident to BAC One-Eleven G-BJRT over Didcot, Oxfordshire on 10 June 1990.* London: Stationery Office Publications Centre, 1992.

Barach, Paul, and Stephen D. Small. "Reporting and Preventing Medical Mishaps." *British Medical Journal* (March 18, 2000): 759.

Brown, David. "Speak Your Mind, But Please Don't Salute." *Navy Times*, May 8, 2000, 20.

Bruno, Frederic. 2000. Telephone interview by author. 20 July.

Carper, Kenneth, ed. *Forensic Engineering: Learning from Failures.* New York: American Society of Civil Engineers, 1986.

Chapanis, Alphonse. *Man-Machine Engineering.* Belmont, Calif.: Wadsworth Publishing, 1965.

Cushing, Steven. *Fatal Words: Communication Clashes and Aircraft Crashes.* Chicago: University of Chicago Press, 1994.

DeQuille, Dan. *The Big Bonanza.* New York: Arlen & Clara Philpott, 1947.

Gerlicher, Scott. 2000. Telephone interview by author. 28 September.

Hennessy, L. W., ed. *Situational Awareness in Complex Systems.* Daytona Beach: Embry-Riddle Aeronautical University, 1994.

Hurst, Ronald, and Leslie R. Hurst, eds. *Pilot Error: The Human Factors.* New York: Jason Aronson, 1982.

LaMendola, Bob. "Pharmacists, Regulators Fear Prescription Mistakes on Rise." Knight-Ridder Tribune News Service, November 2, 1998, K6166.

Loven, Kyle. 2000. Interview by author. 11 July.

Luginbuhl, Michael. 1997. Interview by author. November.

Mathews, Anna Wilde. "Six Seconds, Two Dead: A Police-Van Crash Exposes a Bombshell." *Wall Street Journal*, November 1, 1999, 1.

Maurino, Dan, James Reason, et al. *Beyond Aviation Human Factors.* Brookfield, Vermont: Ashgate Publishing, 1995.

Merrick, George Byron. *Old Times on the Upper Mississippi: The Recollections of a Steamboat Pilot from 1854 to 1863.* St. Paul: Minnesota Historical Society Press, 1987.

Milgram, Stanley. *Obedience to Authority: An Experimental View*. New York: Harper & Row, 1974.

Miller, C. O. 2000. Telephone interview by author. 10 August.

"National Transportation Safety Board Reports on L-1011 Miami Crash." *Aviation Week & Space Technology*, July 30, 1973, 55.

Phillips, Andrew. "The Deadly Price of Oil," *Maclean's*, July 18, 1988, 27.

"Piper Report Blames Occidental and the Department of Energy." *Petroleum Economist* (December 1990): 24.

Reason, James. *Human Error*. Cambridge: Cambridge University Press, 1990.

Rochlin, Gene I. *Trapped in the Net: The Unanticipated Consequences of Computerization*. Princeton, N.J.: Princeton University Press, 1997.

Sandeen, John. 2000. Telephone interview by author. 26 September.

Sinaiko, H. Wallace, ed. *Selected Papers on Human Factors in the Design and Use of Control Systems*. New York: Dover Publications, 1961.

Skrzycki, Cindy. "Risk Takers." *U.S. News & World Report*, January 26, 1987, 60.

Smith, William E. "Screaming Like a Banshee," *Time*, July 18, 1988, 36.

Tullo, Frank. 2000. Personal communication with author. 30 January.

"Unintended Acceleration Mystery Solved," *Road & Track*, February 1988, 52.

Voelker, Rebecca. "'Treat Systems, Not Errors,' Experts Say," *Journal of the American Medical Association* (November 20, 1996): 1537.

Walerius, Chuck. 2000. Interview by author. 29 September.

Young, Bob. 2000. Interview by author. 22 May.

CHAPTER 11

Bryant, William O. *Cahaba Prison and the* Sultana *Disaster*. Tuscaloosa, Ala.: University of Alabama Press, 1990.

Hedges, Chris. "A Key Figure Proves Elusive in a U.S. Suit over Bhopal." *Wall Street Journal*, March 5, 2000, 1.

Kalelkar, Ashok S. "Investigation of Large-Magnitude Incidents: Bhopal as a Case Study." Paper presented to the Institution of Chemical Engineers Conference on Preventing Major Chemical Accidents, May 1988.

King, Ralph. *Safety in the Process Industries*. Oxford, England: Butterworth-Heinemann, 1990.

Kurzman, Dan. *A Killing Wind: Inside Union Carbide and the Bhopal Catastrophe*. New York: McGraw-Hill, 1987.

Potter, Jerry O. *The Sultana Tragedy: America's Greatest Maritime Disaster*. Gretna, La.: Pelican Publishing, 1992.

Salecker, Gene E. *Disaster on the Mississippi*. Wilmington, N.C.: Broadfoot Publishing, 1996.

Shrivastava, Paul. *Bhopal: Anatomy of a Crisis*. Cambridge, Mass.: Ballinger Publishing, 1987.

———. "Long-Term Recovery from the Bhopal Crisis," in *Long Term Recovery from Disasters*, edited by J. K. Mitchell. Tokyo: U.N. University Press, 1994.

Sprick, Thomas. 2000. Personal communication with author. 22 December.

Stix, Gary. "Bhopal: A Tragedy in Waiting," *IEEE Spectrum* (June 1989): 47.

CHAPTER 12

Boyars, Carl. "Troubled Aftermath of an Explosion at Sea." *Chemical and Engineering News*, June 21, 1999, 53.

Cohen, Michael D., James G. March, and Johan P. Olsen. "A Garbage Can Model of Organizational Choice." *Administrative Science Quarterly* (March 1972): 1.

"Damn the Torpedoes. Full Speed Ahead, Dayton," *Surface Warfare* (September–October 1999): 38.

Fledderman, Charles. *Engineering Ethics*. Englewood Cliffs, N.J.: Prentice Hall, 1999.

Furlow, William. "Well Control Shifts to Prevention and Globalization of Response." *Offshore,* September 1997, 108.

Gonsalves, Jerry. 2000. Interview by author. 26 September.

Hoover, R. A. *Forever Flying*. New York: Pocket Books, 1996.

Kletz, Trevor. *Process Plants: A Handbook of Inherently Safer Design*. Philadelphia: Taylor & Francis, 1998.

Kletz, Trevor A. *What Went Wrong?: Case Histories of Process Plant Disasters*. Houston: Gulf Publishing, 1994.

Lewis, Sanford, and Diane Henkels. "Good Neighbor Agreements: A Tool for Social and Environmental Justice." http://www.cpn.org/sections/topics/environment/stories_studies/lewis_henkel.html (Accessed December 15, 2000).

Mills, Don. "Is It Paper or Is It Real?" *Occupational Safety & Health Canada* (April–May 1998): 6.

Mitchell, Pryce. *Deep Water: The Autobiography of a Sea Captain*. Boston: Little, Brown, 1933.

"NASA to Hunt Errors on 172 VA Hospitals; Reporting System Mirrors Aviation's." *Washington Post*, May 31, 2000, A18.

Oliver, Dave N. *Lead On*. Novato, Calif.: Presidio Press, 1992.

Peterson, Walter H. "The Selection of Submarine Commanding Officers." *Journal of Human Performance in Extreme Environments* (September 1998): 13.

Polmar, Norman. *Rickover*. New York: Simon & Schuster, 1982.

Schirra, Walter M. *Schirra's Space*. Boston: Quinlan Press, 1988.

Washburn, Bill. 2001. Telephone interview by author. 25 January.

Wilson, R. E. "Rickover, Excellence and Criticality Safety Programs." *Nuclear Safety* 36, no. 2 (July–December 1995).

INDEX

Aberdeen Testing Center, 247

AeroPeru 757 Flight 603 airliner, crash of, 56, 310

Aerospatiale ATR-42 transport airplane, crash of, 254–55

Airbus airliners, 3, 255

Air France Concorde, crash of, 4–5, 312

Airship Guarantee Company, 69–70

Albin, Fred, 191–92

Alexander, Dick, 194–95

Aloha Airlines 737 airliner, blowout of upper fuselage skin of, 7, 307

Altair Engineering, 184–85

American Airlines, 120–21, 124

American Airlines DC-10 Flight 96 airliner, cargo-door blowout of, 117–18, 123, 125–28, 134, 248, 253, 286, 302

American Society of Mechanical Engineers, Boiler Code of the, 60

American Trans Air DC-10 airliner, oxygen canister fire on, 153

ammonium nitrate, 208, 215–22
  in BASF factory explosion, 218–19, 298
  in Texas City docks explosions, 216–17, 219–21, 299

ammonium perchlorate (AP), 136, 308

Anderson, Warren M., 260–61

Andersonville, prison at, 271

Ansoff, Igor, 4

Apollo 1, 142–50
  casualties of, 142–43, 149, 301
  construction of, 146–47
  oxygen-enriched fire in command module of, 91, 111, 142–45, 147, 149–50, 188, 286, 288–89, 301
  testing of, 147–49, 301

Apollo 7, 278

Apollo 10, 186

Apollo 11, 70

Apollo 12, 183

Apollo 13, 108, 179
  fire in command module of, 188–89
  liquid oxygen tank problem of, 143, 181–83, 185–90
  overheating of oxygen tank in service module external bay of, 70, 143, 183, 187, 189, 302
  thermostat problem of, 186–89, 302
  training for near disasters on, 119–20

Apollo program, 92, 230

Atomic Energy Commission (AEC), 43–44, 101

AT-6 trainer airplanes, communication errors in, 243–44

Audi 5000 sedan, alleged sudden-acceleration problem of, 237–39

Aviation Safety Reporting System, 292

Babcock & Wilcox, 52, 54–55, 61, 119

Badische Anilin und Soda Fabrik (BASF), explosion at, 218–19, 298

Baldwin Hills Dam, collapse of, 190–91, 301

"Ballad of Casey Jones," 257, 296

Banqiao and Shimantan Dams, failure of, 13, 277, 303

Barber, Jonathan, 275–76

barrage-balloons, flying airplanes into cables hung from, 208

Beauvais Cathedral, collapse of choir vaults of, 89

Beech Aircraft, 186

Bell, Arthur, 89–90

Belle Memphis, 270

Bell JetRanger helicopters, 223–24

Bell 206 medevac helicopter, crash of, 144

Benet, Stephen Vincent, 176

Berkun, Mitchell, 177–78

Bhopal disaster:
  casualties in, 259, 261, 263, 266–67, 306
  cause of, 259–60, 263–68, 274, 306
  release of toxic chemicals in, 166, 259–68, 274, 281, 286, 306

BHP Port Kembla Steelworks, damage caused by short circuit at, 141–42

Binks, Joe, 89–90

Black, Ken, 225–26

Black Sea port, explosion of fertilizer-carrying ship at, 221

Blandy, W. H. P., 96
Blue Marlin, 3
Boeing:
    safety and design lessons
        preserved by, 247–48
    707 airliner of, 122
    737 airliner of, 132, 281
    747 airliner of, 141,
        167
    767 airliner of, 112–14
    777 airliner of, 130
Boisjoly, Roger, 65–66, 85,
    88, 91–92
Bolus, Guy, 170, 172–73
Bondarenko, Valentin, 144–45
Booth, Ralph, 82
Bouch, Sir Thomas, 288
Bradbury, Ray, 164–65
Brancker, Sir Sefton, 90
Brescia, Tony, 282
Brest harbor, explosion of
    fertilizer-carrying ship
    at, 221
British Airtours 737 airliner,
    aborted takeoff and
    fire on, 128–30,
    306–7
British Airways BAC 1-11
    airliner, 166–69
    ejection of pilot through
        windscreen of,
        168–69, 309
British Midlands 737 airliner,
    37–39
    crash of, 38–39, 251, 308
Browning, Jackson B., 268
Brownstein, Ron, 278
B-17 bombers, toggle-switch
    confusion in, 243
B-25 bombers, 146
Burrows, Christopher, 104
Butler, Lemmie, 157

Californian, 176
Campolong, Mark, 225–27
Canadian–United States
    Eastern
    Interconnection
    (CANUSE), 184, 301
Capers, Robert, 110
Carper, Ken, 244–45
Carter, Jimmy, 40

Cedar Hill, Tex., collapse of
    broadcast tower in, 137
Cessna Corporation,
    crackstopping
    approach of, 191–92
Chaisson, Eric, 110
Challenger Launch Decision
    (Vaughan), 67
Challenger space shuttle,
    65–67, 112
    casualties of, 88, 91, 307
    crew of, 86–88, 91–92,
        307
    external tank of, 74, 77,
        81, 88
    field joints of, 74, 88–89
    main engines (SSMEs) of,
        74, 87
    midair breakup of, 13,
        66–68, 74, 76–77,
        88–89, 91–93, 97,
        134–36, 138, 286,
        307–8
    origins of, 70
    O-ring seals of, 66–67,
        74, 88–89, 92
    P-12 struts of, 74, 77,
        88–89, 307
    public reaction to, 67–68
    solid rocket boosters
        (SRBs) of, 74, 77, 81,
        87–89, 92, 183, 307
    see also R.101
Chapanis, Andre, 243–44
V. I. Lenin Chernobyl
    Nuclear Power
    Station:
    casualties of, 162, 307
    design problems in
        reactor of, 162–63
    explosions at, 6, 13, 41,
        162–63, 166, 307
    lessons learned from, 163
    power-generation
        experiment at,
        161–63, 307
    safety rules violated at,
        162–63
Cheyenne Mountain
    Complex, 250
Chicago Opera Troupe,
    273–74

Christie, Ralph Waldo, 97–98
Churchill, Winston, 15
Citicorp Center, discovery of
    structural problem in,
    196–99, 291, 303
Clephane, James, 103
cockpit, definitions of, 174
cognitive lock, 41, 175–77
Cole, Cyrus, 279
Cole, USS, 3
Collins, Michael, 289
Colmore, R. B., 83, 86
Columbia space shuttle,
    79–80, 115
D. Colvilles & Company, 85
Comet 1 airliners, blowout of
    fuselage skin of, 7,
    299
Construction Failure (Feld and
    Carper), 244–45
Cooper, Theodore, 279–80
Copeland, Bea, 126–27
copilots, assertiveness of, 251
Corning Glass, 105
Crippen, Robert, 79
Crosby, Joseph, 75–76
Cuban Missile Crisis,
    157–59, 166, 218, 300
Cullen, Lord, 246

Dana-Farber Cancer
    Institute, 255
Darwin, Charles, 132
Daspit, L. R., 95
Davis, John, 200
Davis-Besse Unit 1 reactor,
    55
DC-3 airliner, 122
DC-6 airliner, fire on, 130
DC-8 airliner, 114, 122
DC-9 airliner, surviving fall
    from, 290
DC-10 airliner:
    cargo-door problem of,
        117–18, 122–28, 134,
        248, 286, 303
    initial testing of, 121–22,
        124–26
de Guillebon, Charles, 217
Delta Airlines, 132
Dement, William, 165
Densmore, James, 102–3

Design Objectives and Criteria, 248

"Devil and Daniel Webster, The" (Benet), 176

Dey, Suman, 266

Dietz, David, 43

Discovery space shuttle, 84–85

Dodge, Wagner, 177

Doolittle, Jimmy, 146

Douglas Aircraft, 114, 121–22

Dresser Industries, 52

Dunning, David, 131–32

du Pont, Lammot, 207, 212

DuPont dynamite factory, explosion at, 207, 296

Dyke, Domenic, 27–28, 30, 32–33

Dyno Nobel, 205–7
    nitroglycerine
        manufactured by, 212–15
    runaway chain reactions
        and explosions at, 206–7, 213–14
    safety precautions at, 212–13, 215

Eads Bridge, 279–80

Eastern Airlines L-1011
    Flight 401 airliner,
    crash of, 248–50, 255,
    302–3

Eastman Kodak, 109

Ebeling, Bob, 66

Eggert, William, 125

Emery Worldwide, 153–55

Essex, disaster avoided by, 12

Evans, Cyril, 176

Evans, Dan, 136–37

Exponent Inc., 136

Exxon Valdez, 134, 285, 308–9

Farley, Frank, 256

Faust, Craig, 53–59

Federal Aviation
    Administration
    (FAA), 224
    on Boeing 767, 112–14
    on Boeing 777, 130

DC-10 and, 124
    error reporting
        encouraged by, 292
    oxygen canister fires and,
        151–53

Feld, Jacob, 244–45

Fermi, Enrico, 42–43

Enrico Fermi-1 liquid metal
    fast breeder reactor,
    near disaster at, 108,
    285

Fethers, Rick, 212–14

Feynman, Richard, 85

Firestone tire blowouts, 292

Fitts, Paul, 243

Foecke, Tim, 85

F-100 fighter, 287

Ford Explorers, blowouts of
    Firestone tires on, 292

Foster, John, 288

Frederick, Ed, 53–60

Freeman, E. W., 26

Freud, Sigmund, 56

F-22 fighter, 4–6

Galaxy Airlines Electra II
    airliner, crash of, 9,
    306

Gemini program, 146–47

General Dynamics,
    122–24

General Motors, 104–5, 148

General Public Utilities
    (GPU), 40, 45

Gerlicher, Scott, 230, 233,
    235–37

Gernsback, Hugo, 44

Goddard Space Flight
    Center, 96, 104

Goeben, 15

Grandcamp, fire and explosion
    on, 13–14, 217–20,
    299

Great Oakley, explosion at,
    214

Grebecock, 170

Grissom, Gus, 142, 146,
    148–49

Groening, Matt, 40

Hamilton, Billy, 256–57

Hardy, George, 75

Fanny Harris, 256–57

Hartford Civic Center
    Coliseum, 192–93,
    200
    collapse of space-frame
        roof of, 193, 277, 304

Hartford Steam Boiler and
    Inspection Company,
    141, 185, 274

Hatch, Reuben, 271, 273

Hauss, Clarence, 25–27, 32

Haverfield Inc., 225–26

Hazen, Richard, 154–55

Hermes rocket, 76

Hero of Alexandria, 51

Higgins, Sir John, 82

High Flyer, fire and
    explosion on, 221,
    299

high-voltage transmission
    line maintenance, 9,
    208, 222–28

Hoover, Bob, 287

Hornbeck, Ron, 205–7

hospital-staff mistakes,
    244

Hubble Space Telescope
    (HST):
    cost of, 109
    first repair of, 104, 109,
        305
    investigation on, 105, 110
    misshapen primary
        mirror of, 12, 96,
        104–11, 304–5
    polishing primary mirror
        of, 96, 104–10, 304–5
    press on, 103–4
    reflective null corrector
        (RNC) for, 106–10,
        304–5
    shortcut taken on,
        108–9
    testing and evaluation of,
        96–97, 103, 108,
        110–11

Hubble Wars, The (Chaisson),
    110

Hudson, Jim, 196

Human Rights Watch, 303

Hungarian Carbonic Acid
Producing Company,
explosion of
cryogenic tank at,
1–2, 301–2
Hurricane Alicia, 291
Hylund, Richard, 35–36

IBM, structural problem in
Brazilian plant of,
200–202, 279
Iéna, 133
Illinois Central Railroad,
train wreck of, 257,
296
Institute of Medicine, 244
Instituto Galeazzi, hyperbaric
chamber fire at, 144,
310
Iowa, explosion in gun turret
of, 292

Jarvis, Greg, 87
JCO Tokai Works
Conversion Test
Facility, accidental
nuclear reaction at, 3,
222, 311
John Hancock building,
popping out of glass
windows at, 291
Jones, John Luther "Casey,"
257, 296
Jones, Jon, 144
Jones, R. E., 243
Jones, Richard, 274

Kaminsky, Mrs. Al, 117–18
Kansas City Hyatt Regency
Hotel, collapse of
skywalks of, 11, 305
Kansas City Power & Light,
explosion of
Hawthorn Plant of,
291
Kater, Jim, 132
Katnik, Greg, 80–81
Keenan, Denise, 234–37
Wilson B. Keene, 221
Alexander Keilland, capsizing
of, 19, 191, 305
Kennedy, John F., 218

Kerns, William, 273
Kidd & Company, 136
Kiely, Patrick, 231–34
Kilminster, Joe, 66
Kipling, Rudyard, 49
Kitchener, Lord, 83
Klein, Gary, 189
Kletz, Trevor, 255, 281
Kometik, 12
Kranz, Gene, 111
Kruger, Justin, 131–32
Kubeck, Candalyn, 154–55
Kurtgis, Michael, 224

Lagoona Beach, Mich.,
Fermi breeder reactor
failure in, 108, 285
Lake Superior, survival of
seaman swimming in,
290
Lane, Doug, 224
Lauda Air 767 airliner, loss
of control and crash
of, 14, 113–14, 309
Leech, Harry, 90
Legion Field Stadium,
structural weakness in
upper deck of,
194–96, 300
LeMessurier, William,
197–99, 287
Leslie, Mrs. Shane, 68–69
Liberté, fire and explosion on,
133, 297
Liberty Bell 7, 146
Lipton, Eric, 110
Little, Walton, 150
Arthur D. Little Inc., 267
Lockwood, Charles, 99
Lovell, James, 179
Loven, Kyle, 230–36

McAuliffe, Christa, 86, 91
McCarty, Blake, 234–35, 237
McCormick, Bryce, 118,
120–21, 125–28, 248,
253
McDonald, Allan J., 65–66
MacDonald, Ramsay, 90–91
McDonnell Douglas,
121–22, 124, 128,
150–51, 302

MacLean, Norman, 176–77
McRoy, Wallace, 195–96
McWade, F., 82–83
mail pilots, crash rates of,
280
Maine, explosion of, 12, 296
Malin, Michael, 112
Malmstrom Air Force Base,
157–59
Minuteman 1 ICBM
deployment at,
158–59, 300
"M'Andrew's Hymn"
(Kipling), 49
Manhassett, N.Y.,
construction failure
in, 244–45
Manhattan Project, 42–43
Mann Gulch forest fire,
176–77
Marcot, Jean, 4
Mark 14 torpedo, 95–100
development of, 97–98
failure of, 95–97, 99
Marose, Don, 240
Mars Climate Orbiter, loss
of, 3, 111–12, 225,
311
Marshall Space Flight
Center:
HST and, 105
space shuttle and, 72–75,
84–85
Mars Polar Lander, loss of, 3,
112, 312
Marty, Christian, 4
Mason, J. Cass, 270–73
MD-80 airliners, 150–51
Mehler, Brian, 41, 58–61, 190
Mercury program, 145–47,
149
methyl isocyanate (MIC),
259, 261–67, 281, 306
metriol trinitrate (MTN),
205–6
Meyers, Gerald C., 92
Milgram, Stanley, 251–52,
286
Miller, C. O., 183, 254
Miller, Don, 48–49, 53
Minneapolis Police van
crash, 11, 229–42, 311

automatic shift lock in,
239–41, 311
casualties in, 230, 232,
237, 311
federal investigation of,
233, 235, 237–42
probable sequence in,
232–35, 238–39
Minuteman 1
intercontinental
ballistic missiles
(ICBMs), 76, 158–59,
300
missile attack warning and
response, 6, 253
Mitchell, Pryce, 275–76
Mitsubishi Heavy Industries,
21, 28
Mobil Oil of Canada, 22, 35
Mondale, Walter, 40
Monsanto Chemical
Company, 217,
220
Montagnino, Lucian A., 108
Mont Blanc, explosion of,
218, 297
Morgan, James, 195–96
Morton Thiokol, 65–67,
73–76, 80, 85, 88–89,
91–92
Mulloy, Larry, 66
Murphy, Bill, 12
Murphy, Edward, 11
Murphy Oil Company, 21

National Aeronautics and
Space Administration
(NASA):
and Apollo program, 145,
147–50, 186–87,
230
error reporting
encouraged by,
292–93
faster-better-cheaper
approach of, 164
and HST, 96–97, 103–5,
109–11
and *Mars Climate Orbiter*,
111
and Mercury program,
145–46

probabilistic risk analysis
(PRA) method of,
134
sneak analysis of, 184
and space shuttle
program, 65–67, 70,
72, 75–76, 79–80,
86–87, 91–93, 97,
112, 134
National Highway
Transportation Safety
Administration
(NHTSA), 237–41
National Transportation
Safety Board (NTSB),
183, 254
ValuJet and, 151, 154,
310
Nautilus, 93, 101, 289–90
Navy, U.S.:
Bureau of Ordnance of,
96, 98–99
Bureau of Ships of,
100–102
Development Squadron
12 of, 131
Nelson, Bill, 86–87
New London Independent
School, 156–57
gas explosion at, 157, 298
Newport Torpedo Station,
96–100
New York City:
collapse of scaffold and
hoist mast on
construction site in,
10, 310
sudden-acceleration
disaster in, 236
nitroglycerine, 206–15,
220
discovery of, 209, 211
in DuPont dynamite
factory explosion, 296
dynamite and, 211–15
Dyno Nobel's manufacture
of, 212–15
Nobel's work with, 209–11
Nobel, Alfred, 148, 209–11
Normal Accidents (Perrow),
62
North, G. B., 145

North American Aviation,
287
Apollo project and,
146–47, 149–50, 182,
186–88
Northern States Power
(NSP) Company
building, 230–32,
234–36
Novomesky, John, 200–202,
279
Nuclear Regulatory
Commission (NRC),
40, 52, 55

Occidental Petroleum, 246
Ocean Drilling and
Exploration
Company (ODECO),
21, 25–28, 35–36
Ocean Liberty, fire and
explosion on, 221
Ocean Odyssey, 35
Ocean Ranger, 15, 17–36
casualties of, 34, 131, 306
construction of, 21, 28
crew of, 18–21, 23–34,
130–31, 306
economical use of space
on, 23
emergency brass rods of,
28, 32
emergency evacuation
capabilities of, 23, 29,
31–35, 130–31, 306
flooding of chain locker
of, 33, 35
investigation of sinking
of, 28, 31–32, 35
layout of, 17–19, 22–24
shutting off power of,
31–33, 35
sinking of, 18, 20–21, 28,
31, 35–36, 130–31,
141, 277, 305–6
storms weathered by, 17,
22, 27
suction pump problems
of, 32–33
training of ballast
operators of, 27–28,
32–33, 35–36

(Ocean Ranger, continued)
    as vulnerable to ballasting
       mistakes, 24–25, 36
Olson, Robert, 237
O'Neal Steel, 194
Ophir Mine, emergency in,
    253
ordnance-disposal workers,
    247
Osiris, H.M.S., 289–90

Pacific Engineering &
    Production Company
    of Nevada
    (PEPCON), fire and
    explosion at, 136–37,
    308
Paige, Hilliard, 147
Papin, Denis, 50–52
Pasadena, Tex., explosion at
    Phillips Petroleum
    plant at, 284
Patrick, Joseph, 75
P-delta moment, 291, 296
Perin, Constance, 115
Perkin-Elmer Corporation,
    97, 104–11, 304–5
Perrow, Charles, 62, 282, 285
Petroleine, German
    submarine outrun by,
    228
P-51 Mustang long-range
    fighter, 146
Phillips, John, 175–76
Phillips, Sam, 147
Phillips Petroleum, explosion
    at Pasadena plant of,
    284
Pigott, Jeff, 226–27
Pilâtre de Rozier, Jean-
    François, 6
pilot operated relief valve
    (PORV), 50–54
    origin of, 50–52
    malfunction of, 41, 47,
       52–61, 188, 190, 291,
       304
Pioneer Tunnel, 260
Piper Alpha, explosion and
    fire on, 245–47, 308
Potter, Jerry, 268, 273
Poudre B, 133, 297

pressure cooker, invention of,
    51–52
Prince William Sound, oil
    spill in, 285, 308

Quebec Bridge, collapse of,
    279–80, 296–97
Queen Mary, 21

Rathbun, Donald, 25–30,
    32–33, 35–36
RBMK model power station
    reactors, 163
Reagan, Ronald, 79–80, 91
Richards, Scott, 290
Richmond, Vincent, 71, 77
Rickover, Hyman G.:
    on equipment testing, 93,
       100–102, 280
    and explosion of *Maine,*
       296
    genius of, 278
    Nautilus and, 93
    seven rules of, 283–84
Roberts, James, 2
Roberts, Karlene, 282
Robertson, Leslie E.,
    199–202, 279
Rockwell, Ted, 93
Roddam, 26
R. 100, 70, 82, 91
R. 101, 67–71
    bureaucratic impatience
       with, 82–86, 90, 92
    casualties of, 90–91,
       298
    Challenger compared to,
       see Challenger space
       shuttle
    construction of, 70–71,
       77–78
    court of inquiry on,
       90
    narrative of, 68, 78,
       89–91, 93, 278–79,
       285–86, 298
    fire risk of, 78–79
    at RAF Air Display,
       81–82
    testing of, 78–79, 81–82,
       85–86, 93
Rowena, 270

Royal Aircraft Establishment
    (RAE), 208
Royal Airship Works, 70,
    77–78, 81, 83, 85–86,
    298
Royal Majesty, grounding of,
    3–4, 309

SabreTech Corporation,
    151–55
Safren, Miriam, 244
Sagan, Scott, 158, 283, 300
St. Chad Cathedral, collapse
    of, 67, 295
Salecker, Gene E., 272
Sandia National Laboratory,
    292
Sansone, Robert, 185
Santa Barbara, Calif., oil well
    blowout off coast of,
    285
Sargo, fire in rear
    compartment of, 145,
    299
Sasser, Myron, 194
Saturn rocket, 92, 183
Sawina, Thomas, 229,
    231–34, 236, 240, 242
Schenectady, structural failure
    of, 299
Schiemann, Fred, 53–59
Schirra, Wally, 278
Schwebel, Jim, 11
Scobee, Dick, 87, 92
Sculpin, 279
Seaforth Highlander, 30–31,
    33–34
Seawolf helicopter gunship,
    rotor failure in,
    118–19
SEDCO 706, 20–21, 30
Shea, Joe, 148, 150, 288
Shields, Jill, 290
Shippingport, Pa., nuclear
    power plant at, 44–45
Sholes, Christopher, 102–3
Shrivastava, Paul, 267
Simon, Herbert, 61
Smallwood, Joseph, 145
Smith, Cydya, 118
Smith, Edward, 175
Smith, Marcia, 112

Smith, Morgan, 271, 273
Smith, Peter, 164
Sobrero, Ascanio, 209, 211
Sorensen, Ted, 166
Soule, S. W., 102
Southwest Airlines, safety record of, 281
Soviet Union, 158
  false alerts on attack from, 6, 304–5
  oxygen chamber fire in space program of, 144–45
  Sputnik launched by, 101
space shuttle, space shuttle program, 114–15
  bureaucratic goals for, 67, 79–80, 87, 92–93
  development and testing of, 72–77, 79, 84, 93
  field joints of, 73–75, 77, 84, 88–89, 91
  HST and, 105
  main engines (SSMEs) of, 72–77, 87, 92, 115, 134
  members of public on flights of, 86–87, 91
  return to launch site maneuver of, 115, 119
  R.101 compared to, *see R.101*
  solid rocket boosters (SRBs) of, 72–77, 80–81, 84, 87–89, 92, 115, 135, 307
  superstitions regarding launches of, 135
  see also specific space shuttles
Spartan-Halley observatory, 87, 92
Speed, Frederic, 272–73
Sprick, Thomas, 267
Sputnik, 101
Squalus, 173, 279
Stevens, Joseph, 271
Stewart, Payne, 137, 169, 311–12

Strategic Management (Ansoff), 4
Stromquist, Walter, 131
Stuart, Fla., hospital-staff mistake in, 244
Submarine Force Museum, 289
Sultana, 268–74
  boiler explosion on, 268, 272, 274, 295
Sunjet Aviation Learjet, crash of, 137, 169, 311–12
Swigert, Jack, 188
Szilard, Leo, 42

Tay Bridge, collapse of, 2, 277, 288, 295–96
Taylor, Frederick, 174
Taylor, R. G., 272
Telford, Thomas, 67, 295
Teller, Edward, 44
Tellermine disposal, 247
Ten Commandments of Damage Control, The, 289
Tessenderloo fertilizer plant, explosion at, 219
Texas City docks, explosions of fertilizer-carrying ships at, 13–14, 216–17, 219–21, 299
Thetis, H.M.S., sinking of, 170–73, 280–81, 289–90, 298
Thomas, Rick, 231–34
Thompson, Kent, 25, 31–34
Thomson, Sir Christopher Birdwood, 68, 83–86, 90, 278–79
Three Mile Island Unit 2 (TMI-2), 39–42, 45–50
  cash costs of, 40, 45, 61
  code safety valves of, 60
  construction of, 45
  coolant pipes of, 46–47, 49–50, 53–61
  federal investigation of, 39, 52, 56, 60
  layout of, 45–48
  malfunctioning device in, 41, 47, 52–61, 188, 190, 291, 304

operators of, 46–49, 52–62, 119, 175, 188
partial meltdown of reactor core at, 13, 39–41, 45, 47, 53–62, 119, 166, 174, 190, 283, 286, 291, 304
pressurizer tank of, 47, 50, 53–60
problems with condensate polisher in, 48–49, 53, 57
problems with control room of, 55–57, 59
Rickover on, 283–84
turbine shut down of, 49
Thresher:
  lessons learned from, 140–41
  naval court of inquiry on, 139–41
  sinking of, 140, 159, 191, 300
  testing of, 139, 300
  wartime leak in, 140
Times Square, collapse of scaffold and hoist mast on construction site at, 10, 310
Tinosa, 95
Titanic:
  frustration of radiotelegraph operator of, 175–76
  rivets of, 85–86
  sinking of, 12, 29, 85–86, 268, 290, 297
Titan rocket, 75, 148
TNT, 218, 297
Tonan Maru No. 2, 95
Tripoli, 42
Trojan, 275–77
truckers, 165
Tullo, Frank, 251
Turkish Airlines DC-10 airliner, cargo-door blowout and crash of, 128, 303
Type T personality, 256
typewriter, invention and testing of, 102–3

Ulysses, 87, 92

Union Carbide (India)
    Limited (UCIL)
    Bhopal plant:
    alleged sabotage of, 261,
        265, 267, 274, 306
    maintenance problems at,
        268
    MIC produced and stored
        by, 262–67, 281
    release of toxic chemicals
        by, 166, 259–68, 274,
        281, 286, 306
    safety systems of, 263–64,
        268, 281
Unocal, good neighbor
    agreement signed by,
        284–85
USA Airmobile
    Incorporated, 223–25

ValuJet, 150–55

ValuJet DC-9 Flight 592
    airliner, 150–55
    crash of, 39, 150–51, 155,
        310
Vaughan, Diane, 67
Veille, Paul, 133
Veterans Administration, 293
Vietnam War, 118–19, 224,
        234
Viscount airliner, crash of,
        108–9

Walerius, Chuck, 241
Weick, Karl, 56, 282
Whitesell, Jesse, 213–14
Whitney, Paige (copilot),
        125–28
Wilmore, Carl, 195–96
Winsor, Gordon, 20–21, 27
Winsor, Stephen and Robert,
        20
Wintringer, Nathan, 272, 274

Woods, Frederick, 170–72
World Trade Center, terrorist
    bombing of,
        199
World War I, 15, 69, 228
World War II, 75, 95–101,
        146, 169, 183, 208,
        215–18, 243, 299
    Mark 14 torpedo in,
        95–100
    ordnance disposal in,
        247
    Thresher and, 140
Würgendorf, explosion at,
        214

Young, Bob, 237–42
Young, John W., 67, 79
Young Men and Fire
    (MacLean), 176–77

Zewe, Bill, 53–60